Common Property Resource Management

Common Property Resource Management

A Focus on Forestry

Editors

Amir Ullah Khan
Mousumi Majumdar

ACADEMIC FOUNDATION
NEW DELHI

www.academicfoundation.com

First published in 2011
by

ACADEMIC FOUNDATION
4772-73 / 23 Bharat Ram Road, (23 Ansari Road),
Darya Ganj, New Delhi - 110 002 (India).
Phones : 23245001 / 02 / 03 / 04.
Fax : +91-11-23245005.
E-mail : books@academicfoundation.com
www.academicfoundation.com

Cataloging in Publication Data--DK
Courtesy: D.K. Agencies (P) Ltd. <docinfo@dkagencies.com>

Common property resource management : a focus on forestry / editors, Amir Ullah Khan, Mousumi Majumdar
p. cm.
Contributed articles.
ISBN 13: 9788171888917
ISBN 10: 8171888917

1. Commons--India--Management. 2. Natural resources, Communal--India--Management. 3. Community forests--India. 4. Forest policy--India. 5. Forest management--India. I. Khan, Amir Ulla. II. Majumdar, Mousumi.

DDC 333.20954 22

Typeset by Italics India, New Delhi.
Printed and bound in India.

Contents

About the Editors/Contributors 9

Glossary of Acronyms 17

Foreword 21

Managing Common Property: An Introduction to the Debate 23

I
Forests as Common Property

1. Revolutionising the Approach of Forest Management
   MOUSUMI MAJUMDAR 43

2. Economic Growth, Human Development and Protection of Common Property: Cross-State Analysis of Forest Sector in India
   SACCHIDANANDA MUKHERJEE and DEBASHIS CHAKRABORTY 61

3. Environment and Sustainable Development: New Perspectives, Challenges with a Vision for Future
   ANJALI DEWAN 75

II

Managing Common Property: Case Studies

4. Institutional Issues in Self-Initiated Forest Management Systems of Jharkhand in Eastern India
BHASKAR MITTRA 95

5. Jhum Cultivation: A Threat to Forestry?
MOIRANGTHEM PRAKASH 117

6. Traditional Institutional Mechanisms for the Management of Common Property Resources: A Case From Karnataka
ASHOK S. ALUR, N.V. HIREMATH, R. MOHAMMED, JANE CARTER, RUPA MUKERJI and B.N. DIVAKARA 135

7. Dynamics and Sustainability of Common Pool Resources: A Case Study on the Management of Sacred Groves in Kerala
SUNIL D. SANTHA and K.S. RATHEESHKUMAR 155

8. Evolution from Thengapalli to Community-based Ecotourism: Dhani South Panch Mouja Jungle Surakshya Samiti
PANKAJ SHARMA 167

III

Governance, Laws and Industry

9. E-governance of the Forest: Management by Multivalence
DAVID MICHAEL COOK 177

10. Restoring Indian Forests through Microfinance
K.K. KAUSHAL 205

11. Commons, Community Rights and Statutory Laws: Politics of Forest Use in Uttarakhand Himalayas
PAMPA MUKHERJEE 215

12. Industrialisation of the Bamboo Sector: Challenges and Opportunities
AMIR ULLAH KHAN and ARNAB KUMAR HAZRA 225

13. Agroforestry for Minimising Food, Fodder and Fuel Scarcity
S.K. KARN and SARITA MUTHA JAIN 251

IV

Regional Issues in Forest Management

14. Intra-Community Conflict Resolution Mechanism and the Role of Social Capital: Lessons From a Joint Forest Planning and Management Project in Karnataka
OZMOND ROSHAN D'SOUZA 261

15. Joint Forest Management-led Socioeconomic and Environmental Change in Orissa
R.K. PANDA and C.R. DAS 281

16. Developing Strategies for Rehabilitation of Orans and Gauchars of Western Rajasthan
NAVEEN K. BOHRA and PRADEEP CHAUDHRY 295

17. Forest Dependence and Depletion: Socioeconomic Analysis of Forest Protection Committee Members in the Bankura District of West Bengal
SOUMYENDRA KISHORE DATTA and KRISHANU SARKAR 305

About the Editors/Contributors

Editors:

Amir Ullah Khan is an Electronics and Communication Engineer from Osmania University. He studied at the Institute of Rural Management at Anand, Gujarat and has a PhD in Economics and Business Studies from the Jamia Millia University, Delhi. He is Dean and Director Research at the Bangalore Management Academy, Associate Fellow at the India Development Foundation and Adjunct Professor of Business and Law at the Edith Cowan University. Amir also teaches at the Centre for Executive Education at the Indian School of Business (ISB), Hyderabad and the Indian Institute of Foreign Trade (IIFT), Delhi. He has worked on various research projects for the European Commission, National Council for Applied Economic Research (NCAER), Planning Commission and the World Bank and has written on economics and policy issues. His latest book is titled *The WTO deadlocked: Understanding the Dynamics of International Trade*.

Mousumi Majumdar studied at Visva Bharati University, Santiniketan, West Bengal. She has worked in the field of small scale industries for her PhD in Economics from Visva Bharati University. She has over 10 years of research and teaching experience. She has worked on various research projects for the World Bank as Research Associate in Visva Bharati University, Santiniketan on the environmental research project of 'Common Property Resource Management', then in Indian Statistical Institute, Kolkata on the environmental research project of 'Solid Waste Management'. She has been a Research Associate in Lally School of Management, Rensselaer Polytechnic Institute, Troy, New York on the research project of 'Social Entrepreneurship'. She has also worked as Data Analyst at Bank of America, Menands, New York (USA). She has several conference papers and publications to her credit. Presently, she is a senior faculty at Bangalore Management Academy, Bangalore.

Contributors:

Ashok Sangappa Alur, PhD in agriculture, works for International Crops Research Institute for the Semi Arid Tropics (ICRISAT) as Project Coordinator. He has worked with Swiss Agency for Development and Cooperation, Intercooperation, Agriculture-Man-Ecology and World Neighbors, etc. He also serves as member on Board of Studies, MAU-Maharashtra, and International Climate Adaptation, Global Bio-saline and Climate Change Action Networks, advisor to farmer's federation and International Peasant Movement. Various national and international honors in his credit include 'Honorary Recognition' from Thailand, 'Distinguished Indian' award by Pentagram Research Centre, 'State Honour' from Government of Andhra Pradesh, 'Honorary Credential' and 'SRI Award' from China, 'Basava Jyoti' and 'Scientific Achievement' honours. He has participated and chaired sessions in several national and international conferences and presented papers on international agricultural research and development.

Naveen Kumar Bohra is a Research Officer and Public Relation officer of Arid Forest Research Institute, Jodhpur, Rajasthan. He has PhD in Botany with LLB, LLM, FCC, FDC like many degrees with more than 40 research papers in various journals and about 500 articles in various newspapers and magazines on different topics. He has also 5 chapters in books and presented about 15 papers in different conferences including one international conference organised at Copenhagen in 2009. He has also worked as faculty in FRI University, Dehradun in 2007-08 during his posting at NWFP Division, FRI, Dehradun.

Jane Carter has expertise in natural resource governance and rural development with particular experience in community/collaborative forest management worldwide, but with particular field experience in Asia, community-based watershed development—notably in India, institutional capacity building of community-based organisations focussing on the empowerment of the poor and marginalised, pro-poor and sustainable livelihoods, well-being ranking, gender balanced development, participatory planning, monitoring and evaluation, agroforestry/farm forestry, particularly trees on farms in Nepal. She is currently working as Co-Head Governance and Natural Resources Group for Intercooperation, Bern. She provides back-stopping support to IC projects, and advisory services to others, in governance and natural resources. Own focus on equitable, pro-poor community based natural resource management; continue thematic support to projects in Nepal, Pakistan, Ukraine and Madagascar. Also coordinate IC's membership in the Rights and Resources Initiative (RRI).

Debashis Chakraborty is currently an Assistant Professor of Economics at the Indian Institute of Foreign Trade (IIFT), New Delhi. He has studied at the University of Calcutta and Centre for International Trade and Development, Jawaharlal Nehru University (JNU). His research focus includes international trade policy and WTO negotiations, environmental sustainability and Indian economic development. He has presented and published his research at various academic and policy forums in India and abroad. Before joining IIFT, he had worked at the Rajiv Gandhi Institute for Contemporary Studies (RGICS), New Delhi.

Pradeep Chaudhry is a senior forest officer of Indian Forest Service presently working as Director, State Forest Research Institute, Van Vihar, Chimpu, Itanagar, Arunachal Pradesh. His main research interests include forest and environment valuation, urban parks, gardens, NTFPs, forest management and silviculture. He has many papers to his credit in various reputed national and international scientific journals.

David Michael Cook is a Lecturer at Edith Cowan University, Australia in the School of Computer and Security Science. David has several degrees including a double degree in Asian Studies and International Relations. He also has an MBA from Murdoch University. His main research work has been in the area of governance, where he has specialised in new modes of governance within forestry management. David was formerly the Manager of the SECAU Security Research Centre. He is currently a co-editor of the *Journal of Information Warfare*, and his current research includes a range of governance-related issues specialising in regional issues and developing nations.

Ozmond Roshan D'Souza has a PhD from the University of Hyderabad. He worked on Joint Forest Planning and Management in Karnataka. He is presently working as Research Officer at the Centre for Political Institutions, Governance and Development (CPIGD), Institute for Social and Economic Change (ISEC), Bangalore. His areas of interest are natural resource management (community forestry), climate change, and socioeconomic impact of environmental policies on vulnerable communities.

C.R. Das is presently working as Researcher in Naba Krishna Choudhury Centre for Development Studies, Bhubaneswar, a ICSSR sponsored think tank in Orissa. He has published a number of articles in various research journals of repute. He is associated with a number of research projects in different capacities.

Soumyendra Kishore Datta, PhD in Economics, is a Professor of Economics at Burdwan University for last 20 years. He has supervised 10 PhD works and published about 40 papers. He has done his Post Doctoral studies from University of York, UK, and has completed one major research project on fishery resources, funded by UGC. He has another major project work on forest resources, funded by UGC, which is being continued. Recently he finished a project work on 'Impact Assessment for the Ongoing Government Schemes and Programmes in Birbhum District' funded by the District Magistrate, Birbhum, West Bengal. He has presented several papers in different conferences at national and international level.

Anjali Dewan, PhD in Food and Nutrition, is the Head of the Department of Home Science, St. Bede's College, Shimla. She is interested in the field of nutrition for adolescents, anaemia in women, gender studies, women empowerment, environment conservation, domestic violence etc. She has presented papers on stress management, environment protection and sustainable development at national and international forums. She has been invited as a resource person in various institutes to conduct workshops on nutrition, personality development, career counselling etc. She has submitted a minor project and pursuing a major project awarded by UGC. She was honoured with Woman Achiever's Award, 2010. She writes poems and short stories that are published in leading newspapers.

B.N. Divakara, PhD in Forestry Sciences, presently working as Scientist at Institute of Forest Productivity, Ranchi under Indian Council of Forestry Research and Education (ICFRE), Ministry of Environment and Forest, Government of India.

Arnab Kumar Hazra is a professional economist with Ispat Industries Limited. His specialisation, besides the steel sector, is in law and economics. He has worked and published nationally and internationally in the area of legal reforms, trade, natural resource economics and finance. He has been a staff consultant to the Asian Development Bank, Fellow with the India Development Foundation and also Rajiv Gandhi Institute for Contemporary Studies, New Delhi, Economic Advisor to the Embassy of the Republic of Korea and Consultant to the National Council of Applied Economic Research (NCAER), New Delhi. He was educated at Ram Krishna Mission Residential College (Narendrapur), St. Xavier's College (Kolkata) and Jawaharlal Nehru University (New Delhi).

N.V. Hiremath has PhD in Agriculture Sciences. He is presently working as a senior faculty member for a leading management institute in Bangalore.

Sarita Mutha Jain is a researcher with the Indian Council of Forestry Research and Education, and has worked in socioeconomic and ecological aspects of agroforestry-desertification in arid areas of Indian desert.

S.K. Karn is the Deputy Conservator of forests, presently working with Indian Council of Forestry Research and Education. He has 20 years of experience as forester with expertise in the field of watershed management, agroforestry and joint forest management.

K.K. Kaushal is an Indian Forest Service officer of 1990 batch (Tamil Nadu cadre). He has published his field experiences in various international journals and is invited to several international conferences abroad. He was awarded the Chief Minister's Medal in 2004. He was also awarded the Indian Council for Forestry Research and Education award for excellence for his research on community forestry in 2008 and wrested the Indira Priyadarshini Vrikshamitra Award for the best Forest Development Agency same year as CEO of Harur, FDA.

Bhaskar Mittra has a PhD on community forestry issues in India. He has extensive experience of working in the development sector having worked on issues ranging from watershed management, community forestry, medicinal and aromatic plants and agriculture. He is currently associated with Sir Dorabji Tata Trust.

Raja Mohammed has specialised in ground water management and is presently working as freelance consultant.

Rupa Mukerji is a Post Graduate in Rural Management from Institute of Rural Management (IRMA), Anand. She has vast experience in the areas of natural resource management, social and gender balanced development, stakeholder consultation and implementation and monitoring of projects using participatory approaches. She is the CEO Intercooperation-India (IC India). She also provides advisory services to IC's programmes at the regional level and to global knowledge management mandates. She has worked extensively on issues of rural and urban development, urban poverty, gender and local enterprise development. In the past, she has worked as a consultant on social development, gender and micro-enterprise development to bilateral and multilateral donors, state and Central government agencies, has wide experience in community-based project development. She has field experience in designing and implementing water supply and sanitation, watershed development, rural enterprise and community-based disaster response and mitigation projects in rural and urban India. She is the lead author for over 20 national level consultancy reports and the co-author of 5 recent SDC–IC publications.

Pampa Mukherjee is currently Associate Professor in the Department of Political Science, Panjab University, Chandigarh. She did her MA, M.Phil and PhD from Jawaharlal Nehru University, New Delhi. She specialises in developmental and environmental politics, common property resources, public policy and local level institutions. She has published extensively and also undertaken independent national and international research projects. She has conducted several UGC and ICSSR funded seminars and workshops on quantitative and qualitative methods in social science research. She is currently working on her book on *Uttarakhand: Politics of Identity and Development* (forthcoming).

Sacchidananda Mukherjee is currently Assistant Professor at the National Institute of Public Finance and Policy (NIPFP), New Delhi. He is working on environmental economics, water resources management and public finance related issues in India for last one decade. Before joining NIPFP, he was with International Water Management Institute (IWMI), Hyderabad and World Wide Fund for Nature (WWF)-India, New Delhi. His area of doctoral research is on agricultural non-point source groundwater pollution. He has conducted research projects funded by the Planning Commission, Government of India; IWMI, Colombo and presented and published his research papers at various academic and policy forums in India and abroad.

R.K. Panda is presently working as Senior Research Officer at Centre for Micro Finance Research, Bankers Institute of Rural Development, Lucknow. Prior to this, he was associated with ICFAI National College, as a Faculty in Economics and Quantitative Methods. He was teaching Mathematical Economics and Econometrics at the PG Department of Economics, Khallikote Autonomous College, Berhampur. He has published a number of articles in different journals and presented papers at various national and international seminars. He is associated with a number of research projects undertaken by AFC Ltd., Mumbai and NCAER, New Delhi.

Moirangthem Prakash is an Academic Associate in social science area, Institute of Rural Management (IRMA), Anand, and a member of the editorial team of *Alternative Perspectives*, a quarterly journal published by Centre for Alternative Discourse (CADM), Manipur. He has previously worked at Indian Social Institute, New Delhi and People's Research Society, Bhopal. He is largely interested in understanding the dynamics of conflicts and patterns of development.

Sunil D. Santha is an academic with ardent interests in the field of vulnerability reduction and environmental governance. He currently teaches and leads the research initiatives at the Department of Social Work, Amrita Vishwa Vidyapeetham, Kerala. He has graduated in Social Work from Mahatma Gandhi University and obtained PhD in Sociology from Indian Institute of Technology Madras (Chennai). Presently, he teaches courses related to sustainable livelihoods, environmental change and disaster risk reduction at Amrita University. His research interests are environment and sustainable livelihoods, common pool resources, climate change adaptation, community-based disaster risk reduction and indigeneous knowledge systems.

Krishanu Sarkar is a research student in the Economics Department, Burdwan University. He is currently working as a project fellow in the forestry project funded by UGC. He has experience in surveying about 100 villages in West Bengal. He has published a paper in *Environment and Ecology* journal.

Pankaj Sharma is management graduate from Xavier Institute of Management Bhubaneswar (XIMB), currently working with the Council of Professional Social Workers (Orissa-based CSO) as Project Manager in a community-based Wildlife Conservation and Ecotourism Project. Natural resource management with ecosystem approach is his area of interest.

Ratheeshkumar K.S. is currently pursuing his doctoral studies in Social Work at the Tata Institute of Social Sciences (TISS), Mumbai. His research interests are in the area of environmental health, decentralisation and public health.

Glossary of Acronyms

ACU	Adult Cattle Unit
AD	Avoided Deforestation
APCFM	Andhra Pradesh Community Forest Management
BOJBP	Briksh O Jeevan Bandhu Parishad
CBF	Central Board of Forestry
CBNRM	Community Based Natural Resource Management
CEE	Centre for Environmental Education
CFM	Community Forest Management
CFP	Certified Forest Product
CIDA	Canadian International Development Agency
CIFOR	Centre for International Forestry Research
CIPET	Central Institute of Plastics Engineering and Technology
CO	Carbon monoxide
CPR	Common Property Resources
CPSW	Council of Professional Social Workers
DC	Deputy Commissioner
DCF	Deputy Conservator of Forests
DEG	Digital Era Governance
DFID	Department for International Development
EC	Executive Committee
EC	Executive Committee
EDC	Eco-Development Committees
EKC	Environmental Kuznets Curve

ENVIS	Environment Information System
EQIF	Environmental Quality Index for Forest Sector
ESD	Education for Sustainable Development
EWS	Early Warning System
FAO	Food and Agriculture Organization
FD	Forest Department
FGD	Focus Group Discussion
FPC	Forest Protection Committee
FSC	Forest Stewardship Council
FWI	Landgate Firewatch Indonesia Program
GDP	Gross Domestic Product
GMES	Global Monitoring for the Environment and Security
HDI	Human Development Index
ICAR	Indian Council of Agriculture Research
ICEF	India–Canada Environment Facility
ICT	Information and Communications Technologies
INHERENT	Indonesian Higher Education Research Network
JFM	Joint Forest Management
JFMC	Joint Forest Management Committee
JFPM	Joint Forest Planning and Management
KFDC	Karnataka Forest Department Corporation
LED	Local Employment Days
MDG	Millennium Development Goals
MFP	Minor Forest Produces
MIS	Management Information Systems
MODIS	Moderate Resolution Imaging Spectroradiometer
MoEF	Ministry of Environment and Forests
MPM	Mysore Paper Mills
NCA	National Commission on Agriculture
NEPED	Nagaland Environmental Protection and Economic Development
NERCORMP	North Eastern Region Community Resource Management Project

NHDR	National Human Development Report
NPM	New Public Management
NSSO	National Sample Survey Organisation
NTFP	Non-Timber Forest Products
ODA	Overseas Development Agency
PAPs	Project Affected Persons
PCI	Per Capita Income
PCNSDP	Per Capita Net State Domestic Product
PRIs	Panchayati Raj Institutions
REDD	Reduced Emissions from Deforestation and Degradation
RRA	Rapid Rural Appraisal
RSPO	Roundtable on Sustainable Palm Oil
SCAT	Social Capital Assessment Tool
SHGs	Self-Help Groups
SIFPG	Self-Initiated Forest Protection Groups
SLIP	Shared Land Information Platform
SO_2	Sulphur dioxide
SPM	Suspended Particulate Matter
TE	Total Enrollment
TFP	Tamil Nadu Forestry Project
TNC	Transnational Companies
TVC	Traditional Village Council
UCIL	Uranium Corporation of India Ltd.
UNESCO	United Nations Educational, Scientific and Cultural Organization
VFC	Village Forest Councils
VFDF	Village Forest Development Fund
VFJM	Village Forest Joint Management
VSS	Van Sanrakshan Samiti
WWF	World Wide Fund

Foreword

This book is dedicated to those interested in the environment, and particularly in forests; a group that probably includes all of us. The environment is an intrinsically fascinating subject; exogeneous, extraneous and exotic. It is a strange public good; sometimes financed by taxation, often neglected by the state, captured by vested interests and owned by nobody. Yet it is not that no one is interested in it. Everyone is. The forest is a prime example of common property that is exploited by everyone in the hinterland. Environment like all common property does not have an exclusive title giving ownership rights to someone in particular. Everyone wishes it well, but sees no immediate reward in investing in it or in spending money protecting it.

My interest in common property actually started long ago when my mother used to relate a relatively lesser known fable from her collection of Urdu literature. This one, like several of its kind, was attributed to the inimitable Birbal who cocked a snook, at most of what Akbar believed in. The emperor, happy with his subjects, one day declares that he rules over humble, conscientious and kind people who are caring and concerned about the poor. While the coterie applauds, Birbal is quick to object. In the argument that follows, an experiment is set up.

There is an old empty well that needs to be filled up. The king announces that each subject will contribute milk, at least one glassful and thereby the tank would get filled for the poor to nourish themselves. The king is sure this will succeed, for in his prosperous kingdom, a glass of milk is hardly anything that individuals would grudge. His subjects will in fact pour multiple glasses of milk. 'We shall see', says Birbal. He insists this exercise be carried out at night, so no one gets to see how much the other person has contributed, and there is neither any unhealthy competition, nor an occasion to show off wealth.

As the evening approaches, one clever gentleman puts on his thinking hat. What if he pours a glass of water instead? How will anyone know? Among the thousands of glasses of milk that everyone would pour into the tank, the one glass of water would simply not be noticed. He would be seen going, he would also have marked his attendance and would be seen pouring out his glass in the well. Slyly, he empties his glass of water in the tank. The next morning Akbar goes out in his pomp and glory to inspect the tank. The tank is indeed full, but with thousands of liters of water!

In a classical game theoretic example, this tale tells the story of the environment. No one owns it, so there are no stakes. There is no limit of how much an individual can draw out from the common resource. Private property rights ensure that there is an interest on the part of the owners to maintain what they own in good condition. Common property has no guardian. Or is that so? Why is it that we do see some societies maintain parks better than others? Why are some of the neighbourhoods less polluted than their next door localities? Why is it said that forest dwellers protect jungles in a much better fashion than trained forest guards and officials? How do fishermen ensure a minimum stock of fish in the pond while in other cases all fish are killed in one go? Why does Delhi force its public transport to switch to natural gas while other cities allow smoke-spewing diesel taxis?

— Amir Ullah Khan

Managing Common Property

An Introduction to the Debate

It is often argued that market failures have resulted in the degradation of the environment as individual firms that considered the commons a public good[1] did not have any incentive to protect the environment. In the tragedy that afflicts common spaces, such an argument is often heard. The tragedy of the commons was the term coined by G. Hardin (1968), in his popular paper titled *Tragedy of the Commons*. This rather fancy title refers to the almost certain possibility that any tangible asset that is freely available to everyone is likely to be over-used and eventually exhausted. Faced with such a spectre, governments started intervening and imposing taxes. However, with a complicated relationship within market players, the taxes collected were never efficient in controlling degradation and pollution. This results in governmental failure to enact effective legislation and in effectively enforcing law. Certification bodies and standard setting institutions now have become popular and most contracts and negotiations are carried out between firms that share a common view on standards and have adopted norms that are certified by mutually recognised institutions. For countries like India, it is imperative to focus on enabling market structures to develop sustainable means of addressing environmental issues and concerns regarding climate change. It is also important to understand the nature of the principal-agent relationship in protecting natural resources.[2]

The Kyoto Protocol is an example of an enabling mechanism where emission standards are set and the polluter pays for any extra emission. By allocating the amount of emission each nation is entitled to, the protocol has established systems to ensure that the more developed economies, with large

1. A public good is a good where one person's use does not reduce the amount available for others and where once the good is provided then no one can be excluded from using the good. See Varian (1992).

2. The principal-agent problem is a situation where there is a player (the principal) and one more other players (the agents). How will the agents then work for something that is not in their immediate self-interest? The problem is how to devise incentives which lead to report truthfully and act for the benefit of all.

industries, do not get away with damaging the environment and causing a negative impact on climate change. Also self-regulation by industry aided by various international standard setting bodies and business associations has worked towards setting higher standards all over the world. The core issue we discuss in this book therefore is on how we manage common property and whether the specific concern about forests can be addressed through joint forest management (JFM). It was a happy coincidence that soon after this conference was announced that Elinor Ostrom was given the Nobel Prize (2009) for her analysis of economic governance, especially the commons.

The Royal Swedish Academy of Sciences in its press release[3] said that Elinor Ostrom has demonstrated how common property can be successfully managed by user associations. Economic transactions take place not only in markets, but also within firms, associations, households and agencies. Whereas economic theory has comprehensively illuminated the virtues and limitations of markets, it has traditionally paid less attention to other institutional arrangements. She has challenged the conventional wisdom that common property is poorly managed and should be either regulated by central authorities or privatised. Based on numerous studies of user-managed fish stocks, pastures, woods, lakes, and groundwater basins, Ostrom concludes that the outcomes are more often than not better than predicted by standard theories. She observes that resource users frequently develop sophisticated mechanisms for decision-making and rule enforcement to handle conflicts of interest, and she characterises the rules that promote successful outcomes.

The focus on forests here is because forests have traditionally been used by various groups, both alien and indigeneous. The degradation of forest resources worldwide has led to serious contemplation on the management of common property. If everyone is entitled to graze cattle on a common field, the land is likely to become unfit for other purposes due to over-grazing. Intense study in governing the commons (Ostrom, 1990) has accelerated over the last few decades as this category of problem becomes more critical and pervasive. More generally, common good resource problems share both a public good characteristic—being 'non-excludable' as access cannot be restricted easily—and a private good characteristic as they are 'rivalrous' in nature, that is to say that common resources are 'subtractable' like private property: if one person uses them, another's access is diminished. This is particularly relevant in the case of global warming as the commons is the whole earth system.

In addition, we appear to face a series of tradeoffs that are interrelated: between economic growth and environmental sustainability; between free trade

3. *http://nobelprize.org/nobel_prizes/economics/laureates/2009/press.html*

and equity; between urbanisation and rural development.[4] There are many others. Sitting behind these apparent tradeoffs are three connected challenges: (i) we do not adequately collaborate to understand the problems we face; (ii) the information available to various actors and sectors in addressing these problems is insufficient; and (iii) the sophistication, scale and complexity of our response capacity thus far does not meet what is required. This book and its contributors argue that we can break through these significant barriers if we address these connected challenges with a new approach.

The degradation of the environment has become a serious cause for concern especially after the worldwide debates on this issue since the 1970s. A host of regulation emerged in India, like in other countries as a result of this heightened awareness of the problem. The early approach to regulating the environment came by way of the 'command and control' concept. This form of regulation enacted a law, where clear standards were prescribed and strictly enforced. These standards went so far as to exactly specify what specific technology had to be used by industry. This kind of stringent regulation that specifies technology to be used is most inflexible and on account of this rigidity makes the production process inefficient. A case in point is the Supreme Court insisting that CNG be used for public transport in Delhi. While this did prove beneficial, it is over the long-term that we see the negative impact of legal specification of technology and not the laying down of standards that ties people down. This inefficiency in regulation enabled a debate to emerge questioning the manner in which regulation first came about to prevent environmental degradation. The result was a series of market-based tradable instruments that came about in the 1980s. The search for efficient market-based regulation saw the growth of voluntary standards. Today, firms themselves commit to raise their environmental standards much above the statutory requirement.

Environment is considered a public good, and therefore there is the free rider problem[5] associated with it. Air, water and noise pollutions are example of externalities. The forest, as also all the commons are seen as public goods. In case of public goods, the free rider problem was explained very early in the 20th century by Pigou[6](1928) where he argued for strict government

---

4. According to UNFPA (2007), 'In 2008, the world reaches an invisible but momentous milestone: For the first time in history, more than half its human population, 3.3 billion people, will be living in urban areas.'
5. A free rider is one who does not pay for a good but is still able to consume it because of its public nature.
6. A.C. Pigou was instrumental in developing the theory of externalities. The theory examines cases where some of the costs or benefits of activities 'spill over' onto third parties. When it is a cost that is imposed on third parties, it is called a negative externality. When third parties benefit from an activity in which they are not directly involved, the benefit is called a positive externality. The study of such situations, a part of welfare economics, has been an active area of research since Pigou's efforts early in the 20th century.

regulation explaining that externalities with respect to the abuse of natural resources need strong laws. This is also because the various players involved are intricately tied together in a string of complicated relationships where it is difficult to ascribe costs to each player. Therefore the need to impose taxes and ensure that polluters pay arose. In complete contrast was the market-based system that came by way of the famous Coase theorem.[7] Ronald Coase argued against interventionist welfare activity when he demonstrated that as long as property rights are clearly defined and transaction costs are low, the individuals involved in these situations can always negotiate a solution that internalises any externality.

In India, environment became a concern only in the 1980s.[8] It became a concern only after environment became an economic issue, an issue in world trade and politics. Then suddenly the environment question simply flooded the administration and overwhelmed it by its sheer magnitude and diversity of issues. The policy-setting task, therefore, never quite assumed a central position.[i] The limits of political economy restrict the law operating within. A tradeoff exists between economic growth and environment. But, a developing country like India needs to attain a certain degree of economic growth. This is directly linked to the performance of the private sector. Naturally, insufficient penalties and wide-ranging exceptions and exclusions restrict the full scope of law. After the globalisation of the Indian economy, foreign direct investment (FDI) is perceived to be crucial, and so excessive regulation cannot be a disincentive.

The Indian Constitution and Environmental Law

Specific provisions for the protection and promotion of the environment were absent when the Constitution of India was drafted, debated and adapted by the Constituent Assembly. It was incorporated into the Constitution by the 42nd Amendment Act 1976, in the form of Directive Principles of State Policy (Article 48A) and the Fundamental Duties (Article 51 A (g)). This was in response to the Stockholm Declaration[9] adopted by the International Conference on Human Environment in 1972, which clarifies that man has the

7. Coase (1960) argued that taxes are inherently inefficient and the enforcement causes various difficulties. He offered instead a solution that was based on voluntary contracts and negotiated agreements. This system would do away with governmental intervention and therefore would be better enforced.

8. The immediate trigger was the devastating nature of the Bhopal gas tragedy where human negligence coupled with poor regulation resulted in the killing and mutilation of thousands of people living near a chemical factory.

9. The Declaration of the United Nations Conference on the Human Environment, also known as the Stockholm Declaration, was adopted on June 16, 1972 by the United Nations as the first document in international environmental law to recognise the right to a healthy environment.

fundamental right to freedom, equality and adequate conditions of life in an environment of quality that permits a life of dignity and well being. Article 253 of the Constitution of India states that the Central government may legislate in respect of any subject in the State List, if, it is for the purpose of 'implementing any treaty agreement or convention with any other country or countries or any international conference association or any other body.'

In February 1980, a committee was set up to review the existing environmental legislations, and it was named as the Committee for Environmental Protection, popularly known as the Tiwari Committee.[10] The Tiwari Committee in its review of the environmental legislations had listed out the major shortcomings and had given its recommendations, valid even today.

- Most of the laws were outdated.
- The laws lacked the statement of explicit policy objectives.
- The laws lack adequate provisions for helping the machinery for their implementation.
- The laws were mutually inconsistent.
- There was no procedure for reviewing the efficacy of the existing laws.
- The constitution of a Central Committee to study the laws and suggest suitable reforms.
- The enactment of laws relating to noise pollution and toxic substances.
- The creation of a comprehensive Environmental Code to cover all types of pollution and environmental degradation.
- To empower the Pollution Control Boards with greater powers so that apart from prosecuting the polluters they would also be able to punish them.
- The constitution of Environment Courts in all District Headquarters, and the appointment of experts to assist the Court.
- The creation of a Department of Environment under the charge of the Prime Minister.
- The setting up of a Central Land Commission.
- Elimination of procedural technicalities in pollution cases.
- Amending the laws of evidence so that minor technicalities do not pose hurdles in the way of convicting the polluter.

---

10. The Committee reviewed around 30 statutes relating to the environment, and also reviewed the 200 odd laws, which in the variety of their functions do not virtually protect the environment. The Committee's report was published in September 1980, and accepted by the government soon after.

- Redefinition of the offences under environmental laws and greater emphasis on the compensation to the aggrieved parties.
- Provision of economic incentives to the industries by way of encouraging environment-friendly products, etc.
- Introduction of a law that would mandate that every new industry must issue a public notice about the nature of its activities, effects on the environment, etc.
- Environment Impact Assessment must not only be a prerequisite for the industry to start but also must be repeated periodically.
- Industries should undertake afforestation programmes as part of its social duties.
- There should be only one administrative and implementation machinery and multiplicity of these agencies should be avoided.
- Comprehensive review and reformation of some Central and State Acts, such as those concerning the Insecticides Act, 1968, the Water Pollution Control Act,1927.

Environmental Regulation

The property rights approach, which stems from the Coase theorem has gained wide acceptance in the management of natural resources and even in the field of managing the environment. Good resource management is a function of how well social institutions harness self-interest through individual incentives. This also depends on the information available to self-interested individuals. Parker (2002)[11] has argued that regulatory regimes require organisations to internalise responsibility for their own ethics and regulation through compliance programmes via licence conditions, requirements of enforceable undertakings and other settlements of potential enforcement action, penalty discounts, and even corporate probation orders. Indeed, responsible corporations must self-regulate themselves to prevent, detect and correct breaches of legal and regulatory obligations. Legal obligations have no impact on corporations if corporate decision-makers fail to take them on board as a part of everyday management. Corporate responsibility requires corporate responsibility management. This does not mean that companies can be left

11. In this paper, the conclusion drawn is that evaluation by regulators of corporate compliance management is important because it promotes a virtuous cycle of open self-regulation. The goal is that companies themselves will evaluate their own design, implementation and outcomes of their compliance management systems. It is only through this process of self-evaluation that companies will develop the capacity to detect, prevent and correct their own breaches of ethical, social and legal responsibilities. However they will only be motivated to do so because they know that regulators (and stakeholders) have powerful, sophisticated evaluative capacities to hold them accountable for their attempts at compliance management.

alone to put in place compliance programmes that self-regulate legal and regulatory compliance.

Corporate compliance programmes are only effective when they are open to external stakeholder perspectives and values. The paper goes on to argue that in an ideal situation the open corporation stands for a coming together of government, law and democracy. The need therefore is to encourage and convince corporations to adopt environmental-friendly technologies. In the recent past, this has been done through the focus on corporate social responsibility. Individual or communal property owners who are in a position and have an incentive to obtain time-and-place-specific information about heir natural resource endowments are perhaps better suited than centralised bureaucracies to manage resources. The solution is to be attained by creating a system of checks and balances on both the government and the market. The key is to give the decision-makers the incentives to manage resource properly, by creating appropriate individual or communal property rights.

The Role of the State and Private Sector in Environmental Regulation

The conventional approach to management of resources and environment usually ignores the systems and establishments that provide information to the political sector. Institutions that provide incentives are also largely ignored. Efficiency is of prime importance in the private sector as this has a direct bearing and impact on profits. The whole focus is therefore on getting the information and the incentives in place as it is only then that prices will be correctly set and profits will follow. The public and political sector however functions by the externalisation of costs. These costs are usually borne by the poorer sections of society. Boadway and Wildasin (1984)[12] have shown how a simple cost-benefit analysis helps in quantifying the effect of the use of natural resources. They show that it is possible to measure the change and the economic impact of any shift in the use of resources. In the field of public finance, it can be shown that the cost-benefit analysis concept can be used effectively to calculate the increase in social welfare through specific public policy measures. The use of this tool could effectively provide policymakers with enough evidence to choose between a variety of steps that can be taken in terms of the various methods of protecting the environment.

Product prices are determined by also taking into account the ecological costs as well as the social cost. However, since ecology also involves the

12. The simple point they make for public goods and their management is that people may be unwise, shortsighted or ill-informed.

interaction between different species, it is not easily understood nor can the dynamics of pricing be as simple as the conventional pricing systems. Technology, manufacturing processes and alternative systems also have a direct effect on costs. The government can act by intervening in the market through taxes, financial subsidies and penalties. These measures not only help in the promotion of environment-friendly procedures in both production and consumption but also serve to keep prices in control. Taxation aimed at the initial stages of production and consumption such as taxes imposed on raw material used could be far more effective than taxing the latter stage violators such as taxation of pollutant emitting industries.

International Case Studies on Self-regulation *versus* State Legislation

Haufler (1999) has shown how self regulation by industry aided by various international standard setting bodies and business associations has worked towards setting higher standards all over the world.[13] Various firms have begun to adopt self-regulation to satisfy a large number of stakeholders. This objective of satisfying the stakeholders has resulted in what Haufler refers to as a soft law that is not harshly enforced by the state but is as effective in controlling undesirable behaviour as is constitutional law. Some often quoted examples of international stakeholders ensuring good corporate behaviour in terms of environment protection are Nike and their involvement with international NGOs in setting standards in Indonesia. A number of chemical manufacturers in North America have launched comprehensive care mechanisms that lay down good practices in environmental protection.

Vogel (1996) and Wong (2006) argue that self-regulation is the optimal manner in which firms comply to high standards as this enables them to develop reputation. The reputational capital that is thus developed, out of voluntary practices, pays back in terms of higher share values and larger market shares among a stakeholder base that is increasingly demanding higher standards from firms and industry. A typical example used in this regard is the green funds in Europe that have been increasing in share value as environmental awareness grows.[14] Laws were made that would allow corporate

---

13. Certification bodies and standard setting institutions now have become popular and most contracts and negotiations are carried out between firms that share a common view on standards and have adopted norms that are certified by mutually recognised institutions.

14. Beder (2002) explains how environment started occupying space in the policy arena when the government was educated on how corporate decisions were impacting the environment negatively. The economic reform agenda had started liberalising the industrial policy. This was based on providing free market conditions where each entity would operate freely and the economic considerations would outweigh environmental concerns.

decision-making to focus on maximising profits and therefore allow for increased pollution levels. As a result of protest against this lack of regulation, new laws were enacted and the amount of land under forests had started increasing and the money being spent on environment protection by private business and state agencies had steadily gone up. This combination of government and private action is how a new paradigm in environment protection emerges that relies on both government action and self-regulation.

Globalisation and the Environment

Globalisation has had a major impact on the environment with strong contentions for both its negative and positive influences. For those who advocate that globalisation has had a harmful effect on the environment, proof lies in the poor enforcement of environment protection laws. They assert that increased global trade and foreign direct investments force governments to lower costs which are done by sacrificing environmental regulation and standards. However, on the other hand, those who favour globalisation argue that lower barriers to trade and overseas investment help in the promotion of environment-friendly technology and management systems especially if trade is being encouraged between developed countries with higher environmental standards and therefore better technologies to developing countries where access to such capabilities is limited. They contend that self-regulation has an important role to play in the management and regulation of the environment and that globalisation can greatly impact the need of a firm or business to self-regulate.

Globalisation, by increasing multinational corporation investments in developing countries raises the expectation of their subsidiaries to self-regulate environmental performance in these countries. These multinational corporations can pass on advanced environmental technology to their subsidiaries and therefore enable the transfer of such superior technologies. Faced with competition, domestic firms too will feel pressured to adopt environment-friendly procedures and processes and also self-regulate. Further, globalisation which leads to increased exports from developing countries could impact domestic firms which would feel the need to adhere to strict norms and regulations to appeal to informed and aware consumers in developed global markets where environmental performance could be an important criterion for selection (Christmann and Taylor, 2001). Weidenbaum (2004) looks at the issue of environmental standards and how the pressure on maintaining environmental standards through trade measures is fast gaining ground. However, just as in labour standards, it would be counterproductive to have countries use environmental standards as trade barriers to force foreign firms to comply with domestic norms.[15]

Looking at the Future: The Kyoto Protocol

The Kyoto Protocol of 1997 was a crucial step towards reducing human induced emission of greenhouse gases (GHGs). It has set legally binding emission targets for a basket of six GHGs. Accomplishing these aims would however translate into a basic change in the processes of consumption and production, investment, energy provision and transport in all the signatory countries. Developed countries and economies in transition were committed to a 5.2 per cent reduction in GHG emissions from the levels existing in 1990 by 2008-2012. Issues could arise if governments decide to use subsidies to control and reduce national emission level. Probable subsidies could include the promotion of the use of renewable energy, or the promotion of fuel—switching to sources that are less GHG-intensive in nature. As a result these subsidies would be sector-specific and lead to the impairment of market shares of competing foreign producers leading to conflicts with WTO law.

Do emissions traders exchange 'goods' by the WTO definition? If they are, then the exclusive right to trade them among Annex I parties will violate the WTO's most favoured nation (MFN) principle by discriminating against non-Annex I trading partners.[16] Countries will need to allocate emissions rights to various industries, which can be seen as providing subsidies to certain industries, especially if they are export-oriented, and hence illegal (Cosbey, 1999). The Kyoto Protocol has created an international market for 'carbon currencies'. These currencies or tradable units are of four different kinds, namely: (i) assigned amount units (AAUs); (ii) emission reduction units (ERUs); (iii) certified emission reductions (CERs); and (iv) removal units (RMUs). The United Kingdom introduced a domestic emissions trading scheme in 2002. The EU has been discussing a proposal for an emissions trading scheme. Canada and Japan are also in the process of introducing domestic emissions trading schemes.[17]

---

15. As with any compulsory norm, this too is a measure that would tend to create high barriers to trade and would be distortionary. It would be much wiser to campaign for higher standards and use new techniques to highlight failure in markets to protect the environment. This would be far more effective in actually getting exporting firms in violating contexts to adapt their technologies to environmentally-friendly ones.

16. The MFN principle does not allow any member country of the WTO to discriminate against any other member country of the 150-member strong organisation. If however, emissions credits are 'securities' or transferable financial instruments, then they are not governed by General Agreement on Tariffs and Trade's (GATT) rules. Instead, the institutions that would be needed to oversee the transfer of these securities would have to comply with the General Agreement on Trade in Services, or GATS.

17. Since the ratification of the Kyoto Protocol, trading in carbon rights and domestic schemes for trade in GHG emission allowances are much in use. This trading is what India and in fact South Asia are fast going towards but these would require a certain degree of sophistication in auditing and accounting practices.

The Report on Climate and Trade Rules by the National Board of Trade, Sweden (2004)[18] shows how international trade in emission allowances raises these and other concerns. Can cross-border trade of emission allowances fall under WTO rules? There are a number of interpretations. The non-trade aspect of environment protection therefore is the first. It is but obvious that these allowances are a creation of the Kyoto Protocol and therefore should only be governed by the Climate Convention or the Kyoto Protocol. However the agreement leaves enough room for members to file cases of abuse in the WTO's dispute resolution mechanism. The report showcases how emissions trading is neither a good nor a service. It argues that international emissions trading is the result of governments creating obligations amongst themselves,[19] 'a subsequent rearrangement of these obligations is not a trade in goods or services.'

Under this interpretation, WTO rules would not regulate whether a Kyoto Protocol Annex B party has to allow emissions trading with another Annex B party, a Non-Annex B party or a non-party to the Protocol. It can also be argued that initial allocations are not subsidies. However, what is allocated and traded is an obligation created between governments and therefore is not covered by WTO rules; but regarding a domestic trading systems it is simultaneously argued that allocation of allowances might be viewed as a subsidy (Chambers, 2001). Non-actionable subsidies (Brewer, 2004) are those that cannot be challenged as a basis for undertaking countervailing duties. The Kyoto Protocol authorises the transfer of technology, development of projects, investments and emissions trading between developed and developing countries. The basic feature of trading in emissions as governed by the Kyoto Protocol is an exchange of emissions rights between states.

It is through the Kyoto Protocol that nations are allocated an emissions 'budget' or an 'assigned amount'. These nations then have the right to sell or buy parts of these rights. The Kyoto Protocol does not explicitly require the participation of public and private firms, but it is these entities that will trade in rights. However, it is also important to note that the protocol binds nations and not private firms. Implementation of the rights system is the responsibility of each member state. The protocol sets out, for the first time, provisions for an international emissions trading system, thus, imparting an economic value to each unit of emission reduced and in the process providing a fiscal incentive to find ways to lower the cost of emission reduction on a global basis. The difference in the marginal cost of reducing emissions among the participants

18. It points out that there are three potential areas of concern in the context of WTO rules: (i) the method of allocation of emission allowances; (ii) rules concerning the services that support emissions trading, and (iii) rules for new entrants.

19. See Charnovitz (2003).

is largely expected to optimise the 'returns' on investment by attracting funding to projects where there is a greater scope of emissions reduction per unit of investment. The Kyoto Protocol therefore establishes an enabling mechanism where emission standards are set and the polluter pays for any extra emission. By allocating the amount of emission each nation is entitled to, the protocol has established systems to ensure that the more developed economies, with large industries, do not get away with damaging the environment and causing a negative impact on climate change.

Conclusion: What do the Contributors say?

Historically, market failures have been blamed for having caused the degradation of the environment as individual firms that considered the 'commons' a public good did not have any incentive to protect the environment. That is when governments started intervening and imposing taxes. However, with a complicated relationship within market players, the taxes collected were never efficient in controlling degradation and pollution. This resulted in governmental failure to enact effective legislation and enforcing it. As a result what we see now is a stronger legal institution emerging, but along with market-based mechanisms that allow for private contracts, higher than mandated standards and a movement towards collective action for saving the environment. The debate on this issue is taken forward in this volume. In no way is this a comprehensive solution to the problem. It is not even a complete depiction of all the problems, but is a small step towards debating an issue that remains critical for the political economy in the new world order.

Mousumi Majumdar (Chapter 1) says that the history of forests and forestry in India has changed with the political and demographic scenario. She has described the evolution of forest management in ancient India and how it has come in the modern form of joint forest management (JFM). She highlights how JFM can be viewed as a cooperative game and recommends various steps for its improvement.

Sacchidananda Mukherjee and Debashis Chakraborty (Chapter 2) analysed the underlying factors that determine the performance of a state in protecting its forest resources. Enhancement of human development is generally conducive for better forest related performance for the states, which could be explained by the demand for better environment/conservation with higher awareness. Therefore, the government initiatives for enhancing human development are likely to provide double dividend in terms of positive externalities through environmental benefits as well.

Anjali Dewan (Chapter 3) speaks about the environment and sustainable development. In her paper she discusses the goal of environmental sustainability, which is to minimise environmental degradation and to halt and reverse the processes they lead to. She also touches on the environmental challenges which are to find ways for strengthening the scientific and socioeconomic perspectives, to help authorities make decisions and produce sustainable development strategies with a vision for future.

Bhaskar Mittra (Chapter 4) focuses on the self-initiated community forest management (CFM) systems, which operate in many states like Gujarat, Orissa and Karnataka, and adopts effective forest conservation practices. CFMs have evolved independently, without any assistance from the state-controlled Forest Department, due to acute shortage of biomass and the states' inability to fulfill the basic fire wood, fodder requirements of the affected communities. The case studies at Bero and Kudada in the state of Jharkhand, have explained beyond doubt that how ecological degradation leads to concerns within the community to initiate activities gradually leading to the spread of such measures to surrounding areas and culminating with the formation of a federating body and managed the commendable resurgence of the forest.

Moirangthem Prakash (Chapter 5) has emphasised on Jhum cultivation and deforestation in high altitudes of northeastern region (NER) of India. Over the years, many tribal societies got pushed into employing distorted versions of shifting agriculture. Greater chunks of forest lands have been cleared to grow more food and the return to the fallow plots is much sooner than 10-20 years. This has resulted in deforestation, soil degradation, and fall in yield. It is this change in traditional practice, arising out of the changing conditions, that has given Jhum cultivation a bad name. It is very unfortunate if Jhum only is blamed for denuding of forest in the region. On the contrary, forests are exploited for timber and hills are flattened for soil and stones.

Ashok S. Alur *et al.* (Chapter 6) highlight a case from Karnataka on management of common property resources catering to the needs of seven villages in Kolar district of Karnataka. The paper highlights: (a) inherent strengths of social institutions, (b) effectiveness of practices adopted in water management for crop production and allocation of benefits in terms of social equity, (c) feasibility of continued functioning of institutions, and (d) possibilities for adopting principles, practices and other favourable institutional aspects of traditional institutional structure. These traditional systems may help us in efficient CPR management by drawing lessons from these traditional forms of management especially in matters such as conflict management—to build more effective and sustainable systems of community-based tank management for today.

Sunil D. Santha and K.S. Ratheeshkumar (Chapter 7) argue that the conceptualisation of institution has serious limitations and institutions need not always be functional or purposive, and can evolve even through unconscious interactions/practices of actors. By studying the management of sacred groves in Thazhava village of Kerala, this paper proves that institutions evolve and are continuously shaped by a process through which people consciously and unconsciously draw on existing social and cultural arrangements in response to changing situations.

Pankaj Sharma (Chapter 8) speaks about a *mouja* where a committee generated different mechanisms, included them in their regulating policies and came up with a framework which is even taught to forest officers as a model of local governance and decentralised way of managing common property resources. The committee evolved over a period of time and hence the different approaches—latest being the community-based ecotourism. Thengapalli is a system of local governance which involved physical transfer of wooden stick among different households which is also a symbol of transferring authority and responsibility across. Ecotourism being a noble approach is attracting the youth of village in generating livelihood at the village level through various new livelihood options of working as eco-guides, hospitality and managing camp site etc.

David Michael Cook (Chapter 9) examines whether the new mode of e-governance using cyber, telephone, press, and other electronic media is a worthy method of facilitating participatory involvement in forestry management and then how does it overcome the asymmetrical disadvantages of those regional stakeholders whose livelihoods depend upon a meaningful contribution towards the direction and authority of forestry management. E-governance allows for accurate modelling and good governance in near instantaneous speeds. It reduces the stranglehold of both business and political elites, and permits informal authority to pervade traditional hierarchies of management and power.

K.K. Kaushal (Chapter 10) has highlighted on regeneration of forests through microfinance. This paper presents how the Forest Department of a southern state of India devised a potent tool of microfinance promotion by village forest councils (VFCs) for weaning those who are dependent on the forest while implementing a Japan Bank for International Cooperation (JBIC) funded US $100 million joint forest management project. The paper concludes that the success of JFM is dependent on and is directly linked to the provision of microfinance to villagers through a people's representative body—the VFC. The Forest Department was successful in this unusual task of promoting microfinance even in villages where formal microfinance institutions (MFIs) have failed.

Pampa Mukherjee (Chapter 11) highlights the fact that the introduction of New Joint Forest Management in 1996 has overridden the customary claims of communities on forests, practiced over decades. Further, it shows how boundaries and fences have become sites of anxiety, creating artificial enclaves and plots, and in the process excluding communities from their rightful access to resources in Uttarakhand (Himalayas).

Amir Ullah Khan and Arnab Kumar Hazra (Chapter 12) analysed the challenges and opportunities of the industrialisation of the bamboo sector. Resource management and technical improvements can convert this fast-growing grass into a durable raw material for construction purposes and a wide range of semi-industrialised products. New industrial applications and modern construction design have both demonstrated bamboo's huge potential. In India archaic legislation and lack of awareness have inhibited the bamboo-based industrialisation process. Therefore, what is definitely needed is the linkages that industrialisation of the bamboo sector can provide, given the huge linkages with rural livelihood.

S.K. Karn and Sarita Mutha Jain (Chapter 13) highlighted that during the last two decades the prices of agricultural commodities have in general increased by almost 3 to 4 times but in case of forestry products, e.g., timber, fire wood and charcoal the prices have increased by 15 to 20 times. This change in price structure has made tree growing a highly profitable and competitive land use in some areas. A study was conducted in the western parts of Rajasthan to assess the socioeconomic importance of various traditional agroforesrtry models in arid zones of India. The paper deals with the socioeconomic evaluation of these traditional existing models

Ozmond Roshan D'Souza (Chapter 14) highlights the conflict in benefit sharing among four different communities (hamlets) spread over different geographical areas in a conservation and development programme in Karnataka. It narrates how the managing committee negotiates in the conflict situation with the beneficiaries and how such conflicts can have a negative impact on the conservation activities if there are no policy interventions in unclear cases.

R.K. Panda and C.R. Das (Chapter 15) discuss the issues like delays in the sanction of funds, non–transparency of the Van Sanrakshan Samiti (VSS) functionaries, insufficient provision of training and exposure visits for the members that have partially hindered the progress in the state of Orissa. Considering that there is a tremendous scope for sustainable rural development of the forest fringe households through the JFM programme, the study recommends that all types of constraints associated with such type of governance should be removed.

Naveen K. Bohra and Pradeep Chaudhry (Chapter 16) speaks about the community forests/rangelands called Oran and Gauchar of western Rajasthan. Due to significant economic importance of these areas from time immemorial, local people took special care for protection and sustainable utilisation of these common property resources (CPRs). Salient problems faced by these natural assets of western Rajasthan and strategies for their rehabilitation has been discussed in the paper.

Soumyendra Kishore Datta and Krishanu Sarkar (Chapter 17) emphasise that due to operation of a number of factors like lack of coordination between Forest Department and forest protection committee (FPC), absence of mutual trust among FPC members, extent of forest dependence, availability of alternative income sources, risk of eviction and loss of customary forest access rights, villagers often feel alienated from their traditional property rights and involve in degrading activities of CPR land. Hence grant of permanent tenurial right to forest dependent people, mitigating the fear of eviction, elevating their economic status, extension of alternative employment opportunities and ensuring a cohesive trust between Forest Department and FPC, seem to be the steps that can no more be shelved aside for protecting forest from degradation.

References

Agarwal, Anil (ed.) (1985). *The State of India's Environment: A Second Citizen's Report*. New Delhi: Centre for Science and Environment.

Brewer, T.L. (2004). "The WTO and the Kyoto Protocol Interaction Issues", *Climate Policy* 4.

Bromley, Daniel W. (1991). "Environment and Economy: Property Rights and Public Policy", Oxford: Blackwell Publishers.

Buck, M. and R. Verheyen (2001). *International Trade Law and Climate Change – A Positive Way Forward.*

Carlane, C. (2006). "The Kyoto Protocol and the WTO: Reconciling Tensions Between Free Trade and Environmental Objectives", *Colorado Journal of International Environmental Law and Policy* 17(1): 45.

Chambers, W.B. (2001). "International Trade Law and the Kyoto Protocol: Potential Incompatibilities", in W.B.Chambers (ed.), *Global Climate Governance: Inter-linkages between the Kyoto Protocol and other Multilateral Regimes Inter-linkages*. United Nations University Press.

Charnovitz, S. (2003). "Trade and Climate: Potential Conflicts and Synergies", *Working Draft* 15. Pew Center on Global Climate Change. July. p.2f.

Cheung, S.N.S. (1970). "The Structure of a Contract and the Theory of a Non-Exclusive Resource", *Journal of Law and Economics* 13(1): 49-70.

Christmann, P. and G. Taylor (2001). "Globalization and the Environment: Determinants of Firm Self-Regulation in China", At *http://papers.ssrn.com/paper.taf?abstract_id=277452*. Last accessed on May 21, 2007.

Ciriacy-Wantrup, S.V., and Richard C. Bishop (1975). "Common Property as a Concept in Natural Resource Policy", *Natural Resource Journal* 15(4): 713.

Coase, Ronald H. (1960). "The Problem of Social Cost", *Journal of Law and Economics* 3(1): 1–44

Colchester, Marcus (1992). *Sustaining the Forests: The Community Based Approach in South and South-East Asia.* pp. 69-100.

Cosbey, A. (1999). "Kyoto Protocol and the WTO", *Seminar Note*. International Institute for Sustainable Development. At *http://www.wto.org/english/tratop_e/scm_e/subs_e.htm*. Last Accessed on 7 May, 2007.

Dasgupta, P. and G. Heal (1974). "The Optimal Depletion of Exhaustible Resources", *Review of Economic Studies* 41: 3-28. Symposium on the Economics of Exhaustible Resources.

Demsetz, Harold (1967). "Towards a Theory of Property Rights", *American Economic Review* 57: 347.

Fischer, K. and J. Schot (ed.) (1993). *Environmental Strategies for Industry*. Washington DC: Island Press.

Gadgil, M. and Ramachandra Guha (1992). *This Fissured Land: An Ecological History of India*. New Delhi: Oxford University Press.

Galanter, Marc (1994). *Law and Society in Modern India*. New Delhi: Oxford University Press.

Guha, Ramachandra (1990). "An Early Environmental Debate: The Making of the 1878 Act", *Indian Economic and Social History Review* 27(1).

Hardin, G. (1968). "The tragedy of commons", *Science* 162: 1243-48.

Haufler, V. (1999). "Self-Regulation and Business Norms: Political Risk, Political Activism", in A. Cutler, V. Haufler and T. Porter (eds.), *Private Authority and International Affairs*. Albany: State University of New York Press.

Jinnah, S. (2003). "Emission Trading Under the Kyoto Protocol: NAFTA and WTO Concerns", *Georgetown International Environmental Law Review* 15(1): 709-61.

Lawrence, A., J. Weber and J. Post (2005). *Business and Society: Stakeholders, Ethics, Public Policy* (11th ed.), Boston: McGraw Hill.

Lyon, P.T. and J.W. Maxwell (1999). "Voluntary Approaches to Environmental Regulation: A Survey", *Environmental Economics: Past, Present and the Future*. Hampshire: Ashgate Publishing Ltd.

Mayers, J. and Stephen Bass (1999). "Policy that works for Forests and People", International Institute for Environment and Development, UK.

McKean, M.A. (1986). "Management of Traditional Lands ('Iriachi') in Japan", Proceedings of the Conference on *Common Property Resource Management*. Washington DC: National Research Council. pp.533-589.

Nathan, Dev and Govind Kelkar (2001). "Case for Local Forest Management: Environmental Services, Internalisation of Costs and Markets", *Economic and Political Weekly*, July 28.

National Board of Trade (2004). *Climate and Trade Rules–Harmony or Conflict?*. Stockholm.

North, D. (1990). *Institutions Institutional Change and Economic Performance*. Cambridge: Cambridge University Press.

Ostrom, E. (1990). *Governing the Commons: The Evolution of Institutions for Collective Action*. Cambridge: Cambridge University Press.

Panayotou, T. (1988). "Comments on Kenneth Ruddle's 'The Organization of Traditional Inshore Fishery Management Systems in the Pacific'", in P. Neher, R. Arnason and N. Mollett (eds.), *Rights Based Fishing*. Dordrecht: Kluwer Academic Publishers. pp. 86-93.

Parker, C. (2002). *The Open Corporation*. Cambridge: Cambridge University Press.

Pearce, D.W. (1941). *Economics and Environment: Essays on Ecological Economics and Sustainable Development*. Cheltenham: Elgar Publications.

Petsonk, A. (1999). "The Kyoto Protocol and the WTO: Integrating Greenhouse Gas Emissions Allowance Trading into the Global Market Place", *Duke Environmental Law and Policy Forum* 10(1): 185-220.

Pigou, A.C. (1928). "A Study in Public Finance", *The Economic Journal*. London: McMillan.

Repetto, Robert (1997). "Macroeconomic Policies and Deforestation", in Partha Dasgupta and Karl-Goran Maler (eds.), *The Environment and Emerging Development Issues* 2. Oxford: Clarendon Press.

Runge, C.F. (1986). "Common Property and Collective Action in Economic Development", *World Development* 14(5): 623-35.

Sarin, Madhu (1996). "Who Is Gaining? Who Is Losing? Gender and Equity Concerns in Joint Forest Management", *Working Paper* by the Gender and Equity Sub-Group, National Support Group for JFM. New Delhi: Society for Wasteland Development.

Schwartz, Herman M. (1958). *States versus Markets: History, Geography, and the Development of the International Political Economy*. New York: St. Martin's Press.

Singh, Gurdip (1995). *Environmental Law*. New Delhi: Lawman (India) Private Limited. p. 185.

UNFPA (2007). *State of World Population 2007*. (Introduction).

Varian, Hal (1992). *Microeconomics Analysis*. New York: W.W. Norton & Company.

Weidenbaum, M. (2004). *Business and Government in the Global Marketplace*, (7th Ed.). New York: Prentice Hall.

Wiser, G.M. (1999). "The Clean Development Mechanism versus the World Trade Organizaion: Can Free-Market Greenhouse Gas Emissions Abatement Survive Free Trade?", *Georgetown International Environmental Law Review* 11(3): 531-598.

Wong, S. (2006). *Environmental Technology Development in Liberal and Coordinated Market Economies: Tweaking Institutions*. New York: Palgrave Macmillan.

I

Forests as Common Property

1

Revolutionising the Approach of Forest Management

MOUSUMI MAJUMDAR

The history of forests and forestry in India has changed with the political and demographic scenario. With the march of civilisation, the population increased at a very high rate and this began to have an adverse effect on the forest ecosystem. In the next stage of development, when the human race came under the false notion that it could conquer nature, reckless exploitation of forests took place. The impact of human activities created a serious threat to overall ecological balance between man and the environment.

Ancient India Scenario

Forests in ancient India were an integral part of people's lives. The association of the Indian people with the forests was so intimate that Indian culture was known as *aranya samskriti* and forests as *aranyani* (Bandhyopadhyaya and Shiva, 1987). Humans have lived as hunter-gatherers over most of their evolutionary history and therefore the social organisations of the primitive societies were centred around their dependence on nature. While no formal management of resources existed, the ancient societies exerted effective communal control of the resource exploitation by evolving many cultural restraints, which included days and seasons of abstinence from hunting and/or gathering, prohibition or killing some sacred species of plants and animals, protection of sacred groves dedicated to some deities, etc.

As reflected from our rich heritage of myths and legends, the relationship between the people and the forests was one of love and reverence, exemplified by the planting of species whose span of maturity exceeded human lifetime. Such acts were sincerely performed from the 'sense of responsibility towards future generations' (Haimendorf, 1955).

The agricultural and pastoral society that followed the hunter-gatherer societies adopted many cultural practices of using forests from the tribals and thus incorporated the tribal view into the Hindu *vedantic* philosophy. Thus, religion and folklore constructed a protective ring around the forests on which the people living in the surrounding settlements depended for the grazing of their cattle, obtaining material for their crafts, and for producing food, fuel and medicines. Competition for the resources was reduced through diversification of resource by different occupational caste groups.

The *Vedas*, the *Upanishads*, and the *Puranas* reveal the vital role of forests and forestry during those ancient days. The epics *Ramayana* and *Mahabharata* give an attractive description of forests like Dandakaranya, Nandavana and Khandavana and the ancient Hindu culture is said to have its origins in the *Aranyas* (forests). Apart from these natural forests, there is a mention of artificial forests such as Amaravana, Devadaruvana, Shirishavana, and Osadhivana.

Traditional Indian lifestyle evolved in a manner compatible to the protection of environment and conservation of forests and wildlife. Ancient religious philosophy and customs of nature worship, bear the testimony to the fact that traditions laid down by the religious and social leaders of those times were designed to be eco-friendly.

Evidence of early forest management by the state is found in Kautilya's *Arthasastra* (BC 321), during the reign of Chandragupta Maurya, which refers to 'forests' being managed as state reserves for revenue and *lokvan* for public use. Forest conservation measures were initiated by Emperor Ashoka, as is testified by the decrees inscribed in the rock and pillar edicts. Much importance was given to planting trees along the roads and on camping sites. This concern continued till the beginning of the 6th century. In the Mughal period, there did not seem to be any interest for the conservation of forests. However, Akbar showed some interest in planting trees along canals and highways. A few species of trees were valued by the rulers and declared as 'Royal trees'. Otherwise, forests were open to all and the people met their requirements of forests produce without any restriction. There was a gradual destruction of forests. Unregulated clearance of forest area for agriculture and overexploitation of trees for timber had adverse effects on India's forests.

British India: First Phase

The exploitation of forests was particularly severe during the later part of 18th century and the first half of the 19th century, especially in the accessible areas. It was overexploitation and degradation of teak forests along the Malabar Coast to meet the requirements of British navy, which brought into focus the need to regenerate the forests and conserve the timber sources. This led to the establishment of Forest Departments in several states. In 1800, a commission was appointed to enquire the availability of teak in the forests of Malabar. Regulations followed, prohibiting felling of young trees. In 1805, a forest Committee was constituted to enquire into the capacity of forests. In 1806, a conservator of forests was appointed in Madras to organise timber supplies from the west coast. In 1855, Lord Dalhousie, the then Governor-General of India issued a memorandum termed 'Charter of the Indian Forests' which outlined the rules for conservation of forests in the whole country.

British India: Second Phase

Scientific Forest Management Policy and Legislation

Systematic management of forests in the country began in 1864 with the appointment of Dietrich Brandis, a trained German forester, as the Inspector General of forests. Simultaneously with the advent of scientific management of forests, the government decided to treat forests as state property by enacting the Indian Forest Act, 1865. This act was the first attempt at forest legislation in India. Between 1869 and 1878, while drafting forest laws for different provinces, it was ensured that the long-standing customary rights of people were protected. This act was revised in 1878 and provided for the constitution of reserved forests—where most of the customary rights were extinguished—and protected forests—where the customary rights were allowed to continue. By 1882, all the areas falling under the British rule were covered by the Indian Forest Act. The first Forest Policy of the country was formulated in 1894 on the basis of a report by J.A. Voelcker on the state of agriculture in India. In 1927, the act of 1878 was consolidated to regulate the law relating to forest and forest produce.

Regulations of 1878 and 1894 negated the tribal conviction that forests belong to them (Elwin, 1963) and reasserted the full authority of the state. The 1878 Act sharply defined and delimited the amount of forest produce each family was allowed to collect from the reserved forests. The villagers had the rights of use over the protected forests, also under the state management. They could hunt and gather all forest produce for domestic use (but not for sale), and could also graze their livestock in these protected forests. The relatively 'unproductive' village forests were left for the villagers as their communal property.

The demarcation of the reserved forests aroused suspicion among the villagers that any trouble they may exercise in improving their village forests would end in taking away of the forests by the Forest Department as soon as they became commercially valuable (Ribbentrop, 1900). In many cases, the village forests were ruined by the villagers for fear of appropriation by Forest Department (Guha, 1989).

During the first quarter of 20th century, scientific aspects of forestry began to receive greater attention. The Imperial Forest Research Institute was established in 1906. Chief conservators of forest began to be appointed for each province in 1907 and in 1910, a board of Forestry, composed of conservators under the chairmanship of the Inspector General of Forests was set up. Re-enactment of Indian Forest Act in 1927 gave further impetus to forest protection and management.

The gains of the policy and administration of forest conservation received a severe setback during the two world wars. The wars accentuated overfelling and most of the forests were destroyed.

Forestry Development: Post-Independence Scenario

After independence in 1947, the erstwhile princely states with their vast lands were integrated into the country and large private forest areas held by landlords (*zamindars*) came to be vested with the Government. Thus, more areas were brought under scientific management and laws relating to forests were unified. In order to aid and advise the Central and state governments/ union territories in all matters concerning forest conservation, management and related matters, the Central board of Forestry (CBF) was constituted in 1950. Amidst the various changes, the National Plan of Development commenced in 1951. In 1952, a new National Forest Policy was enunciated which proposed classification of forests on functional basis into protected forests, national forests, village forests and tree lands. Plantation and afforestation have been the major activities in the forestry sector since the beginning of the First Five-Year Plan. Under the five-year plans, artificial man-made forests were begun to be created on an unprecedented scale. Concerted efforts triggered off increasing productivity of all forest sites maximising the wood resources. Thus, forestry's development found an important place in national plans.

A National Commission of Agriculture (1970) was appointed by the Government of India, which advocated an ambitious programme of production and social forestry and called for the creation of large scale manmade forests with the help of institutional finance. It also suggested institutional changes in the management of forests for production forestry.

Centralised control over forest management without due weightage to people's perceptions had the following effects:

1. The local communities began to have a diminishing access to the forest resources, which provided them with several basic necessities such as fuel, fodder and food. This had serious effects on their lifestyles and socioeconomy.
2. Traditional practices and mechanisms, which regulated the local use of forests through a set of time-tested practices, broke down.

It has now been realised that centralised control over forests without enlisting the cooperation and support of the local people could not prevent the degradation of forests, which has continued over the past centuries and does so now, though at a relatively slower rate. It has eventually been realised that without the willing and active participation of people living in and around the forests, it would be difficult for any programme of sustainable forest management to succeed.

In early 1970s, however, experiments in joint forest management (JFM) were initiated, which led to a new era of forest co-management. JFM initiatives are based on the concept of collaboration between local people and state authorities. Local people participate in forestry activities on land that remains, essentially, under state control; the Forest Department provides financial assistance and technical advice. JFM have been successful in various parts of India and also in West Bengal. But the efficient functioning of the forest protection committees depends on, in many cases, the personal efficiency and willingness of concerned Forest Department officials.

Far-reaching developments in the demographic, economic, social and environmental fields since 1952 resulted in the revision of National Forest Policy again in 1988. The new policy marks a major departure from the earlier one of 1952, which was mainly production-oriented. The current policy accords highest priority to ensuring environmental stability and maintenance of ecological balance. The national goal is to have a minimum of one-third of the total land area under forest or tree cover.

In the present situation, with decaying traditional management on one hand and totally unsuccessful centralised management on the other, government is gradually opting for co-management in the name of the policy of the JFM. JFM is a participatory approach and is basically a contract between two groups of players—the government and the villagers, success of which depends on the relative strength of each party and the relative importance of the resource at stake.

India's JFM initiatives are based on the concept of collaboration between local people and state authorities. The idea was to provide residents with a supply of biomass and sources of income through the sale of non-timber products like fruits, leaves, mushrooms, twigs and fodder grass—and in exchange the communities would help to restore and protect the forests.

Thus, three types of local level common management are conceived:

a) *Traditional management*: where local level users were fairly successful in conserving the natural resource at stake.

b) *Centrally managed system*: where resource management is basically done by some Central authority like department of forest, under the state as well as the Union government.

c) *Co-management*: where the relationship between resource users and governing authority is one of partnership.

Concerned officials realised that poor villagers could not be restricted from forests; instead, alternative livelihood methods were needed to stop the unsustainable exploitation of forests. It was out of these concerns that the two pilot experiments of the early 1970s—in Arabari in West Bengal and Sukhomajri in Haryana—were introduced.

In the Arabari range of Midnapore district in West Bengal, the local villagers were motivated to form forest protection committees (FPCs). These were to protect public forests in return for first preference in forestry department employment, free usufruct rights to most non-timber forest products (NTFPs) and a 25 per cent share in the net cash benefits from sales of short rotation timber poles. An initial success led to a gradual expansion of the efforts to neighbouring areas and increasing support from West Bengal's forestry department. The concept of 'social fencing', which evolved in Sukhomajri in Haryana, focused on enabling the villagers to reduce their dependence of unsustainable grazing practices in fragile, forested hills. However, Haryana's programme remained small, while the West Bengal JFM framework expanded dramatically and served as a model for other states.

The new national forest policy of 1988 reversed the states' earlier forest management priorities of meeting industrial and commercial requirements for forest produce and maximising revenue. Instead it emphasised on protection, conservation and meeting the needs of the rural poor thereby laying the ground for JFM (Government of India, 1988). This was followed on 1 June 1990 by a circular (No. 6.21/89-FP) from the Ministry of Environment and Forests to the forest secretaries of all states and union territories providing guidelines for the 'Involvement of village communities and voluntary agencies in the regeneration of degraded forests'.

To date, 15 state governments have issued their own resolutions (government orders) specifying the basis of JFM partnerships with local communities. These 15 states have 74.6 per cent of the country's 75 million ha of public forest land and 91.4 per cent of the country's tribal population. As of the latest information compiled by JFM cell in the MoEF, as on 10 September, 2003, 17.33 million hectares of forests are being managed through 84,642 JFM committees in 27 states (JFM Cell, MoEF, 2003).

Although precise information is scarce, there are several indicators of the positive impact that JFM has had on forest conditions (Sarin, 1998). In many states, forests under this form of management are regenerating with remarkable vigour and diversity. Local community institutions are protecting their forests far more effectively than the state Forest Departments could. West Bengal has the largest programme with over 2,300 registered FPCs protecting approximately 3,20,000 ha of public forest land.

A number of studies, therefore, emphasise the link between poverty and environmentally sustainable agriculture. Chambers (1987) argues that sustainable development can be attained through positive impact on the livelihood of the poor. Vyas (1991) thinks that properly conceived poverty alleviation programmes could be a step in the direction of an environmentally safe world. According to Nadkarni (1993) poverty alleviation programmes that ignore the dependence of the poor on common property resources (CPRs) are likely to be abortive, and an integrated view of development must be taken for both poverty alleviation and environmental protection.

Thus, the formulation of the appropriate CPR management regime for a particular context assumes importance (see for example Singh, 1994; Noronha, 1997). Prescriptions range between state ownership or intervention (Hardin, 1968) and privatisation (Hardin, 1978; Demsetz, 1967), both of that are controversial. In particular, Chambers *et al.* (1990) notes that state monopoly often ends up depriving the poorer sections of the benefits of CPRs and ultimately alienating them. Nadkarni *et al.* (1994) found that the villagers stopped cooperating with the Forest Department when they started taking up village grazing land for social forestry. Similarly, the problems of privatising CPRs, by way of land reforms or by providing incentives to higher income groups have been widely discussed in the literature. Binswagner (1989) has shown how acquisition of forest lands by the rich in Brazil led to deforestation.

When appropriate consideration is taken regarding the livelihood of the poor, however, there is much greater success in the outcome. Thus, Singh and Bhattacharjee (1991a) find, in an in-depth case study of Nepura village in the Midnapore district of West Bengal, that the process of State-led privatisation of CPRs through land reforms, when formulated with due considerations to the interests of the poor, can be an important instrument in achieving the desired objectives of equity, efficiency and sustainability in CPR management.

It is in this context the importance of participatory management of CPRs has been highlighted (Singh, 1994). This could take the form of cooperatives in fishing (Singh and Bhattacharya, 1991b), tree growing in forests (Singh and Ballabh, 1989). Village *panchayats* may also be quite effective (Brara, 1987), although it is argued that excessive politicisation and opportunism often prevent them from enforcing user regulations, turning CPRs into 'open access resources' (Gupta, 1987; Bromley and Cernea, 1989).

A Study of West Bengal's FPCs (Indian Environmental Society/Ford Foundation, 1990) claims that groups with at least 75 per cent of all communities participating are the most effective forest management units; tribal communities and groups with heavy forest resource dependencies make better forest managers than less forest dependent people.

Another study carried out in the district of Bharuch (Tiwari, 1994) in Gujarat suggests that the following factors contribute to the success of collective protection of forest resources:

(i) size of membership of Gram Vikas Mandal (GVM)/Van Kalyan Samiti (VKS);

(ii) size of total illiterate population;

(iii) size of expected economic benefits for JFM;

(iv) number of visits by the coordinating agency; and

(v) years of interaction with the coordinating agency.

The following features contribute negatively:

(i) the size of village;

(ii) unemployed graduates of the village; and

(iii) livestock population of the village.

In the theoretical literature regarding participatory management of commons, including forest, Ostrom (1999) finds that the following characteristics contribute to the success of such institutions:

(i) participation by those most affected;

(ii) monitors/supervisors accountable to participants;

(iii) sanctions/fines are graduated (i.e., proportional to offences);

(iv) rights not challenged by authority; and

(v) transparent rules and easy dispute resolution.

Briefly, the pre-state control phase is one of unrestricted access that can be modelled as a 'Prisoners' Dilemma' game in which each individual extracts more than what would be jointly beneficial. By contrast, the situation under state control is one of complete loss of property rights by the local villagers. The State here maximises the difference between (long run) revenue from the sale of timber and the expenditure on forest protection. Given such expenditure, which determines the number of officials, the villagers then maximise the difference between gains from theft and (probable) penalties if they are caught.

In the context of the real success of JFM, the interesting issue is to demonstrate that sharing a part of timber revenue to the villagers and giving them access to NTFP in exchange for forest protection is profitable for the government. Where successful, such cooperation can reduce: (a) the protection expenses of the state, and (b) theft of forest produce, sufficiently to pay for its lower share of revenue. A further issue is to examine how this success depends on economic, social and institutional factors.

Among the likely determinants, clearly the dependence of the participants on the resource and the income earned from it are likely to be important factors. So is the level of penalties from theft, as is perhaps the rewards from preventing it. We shall attempt to include these factors in our theoretical model.

Joint Forest Management (JFM) in India

India's JFM initiatives are based on the concept of collaboration between local people and state authorities. Local people participate in forestry activities on land that remains, essentially, under state control; the Forest Department provides financial assistance and technical advice. JFM grew out of tension in 1970s and 1980s between Forest Department staff and local communities. This was an era of political upheaval in many states. Villages had increasing need for forest resources but decreasing access to them, as government aggressively promoted state plantations in barren and degraded forest lands that had always been used by local people. During this period, in 1971, Dr. Ajit Banerjee, a young divisional forest officer, Midnapore district, West Bengal, launched a most innovative project called 'Socio-Economic Project' in a cluster of 11 villages in the Arabari block of the district in which local villagers worked with Forest Department staffs to jointly manage forest patches adjacent to their settlement. The idea was to provide residents with a supply of biomass and sources of income through the sale of NTFPs—fruits, leaves, mushrooms, twigs and fodder grass; first preference for employment; 25 per cent share in the net cash benefits from sale of short rotation of timber poles—and in exchange communities would help to restore and protect the forests. Soon, 618 families from 11 villages started working with West Bengal Forest Department to restore more than 1,272 hectares of forest. By early 1980s, jointly managed forests in Arabari were flourishing. Today 'Arabari experiment' has been formally endorsed as a general model for jointly managing forests.

There is a growing view across the world that without strategies such as co-management, collaborative management and joint management, the forest conservation would not be possible. In India, the official blessing for such an approach originated in the National Forest Policy, 1988 which recognised the interdependence between people and forests which envisages active community participation in the protection and development of forest lands. JFM is interpreted as the implementation of this policy. Many of the states have obtained foreign aid to support JFM on their respective forestlands. It has been estimated that nearly 10,000 communities have formed FPCs that can cover 1.5 million hectares of forest. JFM is however merely an administrative arrangement and does not have the sanction of the Indian Forest Act, 1927 which is still the main law governing forests in India.

Along with decentralisation, however, there is an element of centralisation that seems to be introduced as a balancing mechanism in the structural composition of the executive committee (EC) of the FPC. The inclusion of head of the local *panchayat samiti* and the *gram pradhan* as ex-officio members of the EC introduces an element of state control. Here, perhaps lies the crux of the problem of politicisation of an apolitical structure. The FPC is essentially a non-political structure. Like the FPCs, the Forest Department too is a politically neutral organisation in as much as it is a part of the bureaucracy and the administrative setup.

Membership in the FPC is restricted to one representative from each household. It could either be a male or a female. Female membership ranges from 5 per cent to as much as 63 per cent. However, in most cases male representation invariably seems to be the norm.

JFM as a Cooperative Game

JFM is a paradigm shift in the sense that it makes a departure from the historical policing approach to one of seeking community support for forest conservation. JFM is a desirable institutional alternative which lies somewhere between state-controlled forest management model and the 'free for all' model. The symbiotic relationship between rural people and their surroundings, from which they draw sustenance, is considered central to the newly emerging concept of natural resource management.

JFM calls for a radical change in the attitude of the staff towards the local communities. A logical consequence of this is the need for reorientation, training and capacity building in the forest staff. In the game theoretic terminology the JFM can be called the win-win strategy, i.e., both the Forest Department and the local community gain from this approach and the forests are conserved. To increase the mutual cooperation among the people and the Forest Department, there must be frequent interaction. The tit-for-tat policy (i.e., one party takes the approach of cooperation so long the other co-operates: Axelrod, 1984) pays off if the chances for interaction in future are high. In such a situation the tendency would be to cooperate rather than mutual antagonism.

The Win-Win Strategy

Relationship of the people and Forest Department can perhaps be abstracted as one of Prisoners' Dilemma, a classic example in game theory to illustrate the two person non-zero sum game. The description of the game is as follows:

The game structure in the Prisoners' Dilemma is such that the dominated strategy for both the players is to confess, which would end up in making both the parties losers. If one decides to cooperate, i.e., not confess and other confesses, the prisoner who cooperates ends up being the loser. To cooperate or not is the dilemma.

The same game can now be depicted in the context of forest management with two actors, i.e., the Forest Department and the people.

Table 1.1

Forest Department and the People

Forest Department	*Village People*	
	Cooperate	*Not Cooperate*
Protect with people	Both gain. Forest conserved. Forest management strategy.	It is what happened in social forestry and the programme did not succeed.
Protect without people	Forest protection by people but not recognised.	Forest degradation as it happened in the past.

The pay offs are shown in each of the cells for the strategies. Thus we can see that the non-cooperative alternative has cost both the Forest Department and the people dearly and has also resulted in the loss of forests. JFM as a co-operative strategy seems to be the best alternative not only to recover forest but also to gain in the process.

Joint Forest Management (JFM): The Organisational Structure

In JFM, the Forest Department has set up its rules, which it hopes that people will find worthwhile to follow. JFM hinges on the crucial sharing of the forest resources between the people and the Forest Department. JFM is all about how the principals organise themselves in the presence of the external agent whose presence cannot be washed away. Further, the external agent is also imposing rules that are not necessarily favouring the principles. There is also the phenomenon that the communities have organised them as though there is no external agent (or its influence is marginal) and it is widespread in the eastern part of India. The participation of the people in the context of forestry has led to the emergence of tree growers' cooperatives, village FPCs set up by the state Forest Department, and self-initiated forest protection groups and variations, hence referred to as people's institutions (PIs). PIs are used to denote cooperatives in Gujarat and FPCs in West Bengal and Orissa. It is estimated that about 10,000 PIs are involved in the protection and conservation of forests. Majority of them are considered to be self-initiated and this phenomenon is particularly evident in the states of Orissa, Bihar, Gujarat and West Bengal. In West Bengal there is, to some extent, merger of the official and the informal. These PIs have responded to the growing deforestation and the resulting short supply of forest resources essential to them by organising themselves to form forest protection institutions. They have set up their own rules, their own committees, defined the boundaries of the forest that they are protecting and have helped to regenerate about 1.5 million hectares of forests.

Figure 1.1

Structure of Forest Protection Committees (FPCs)

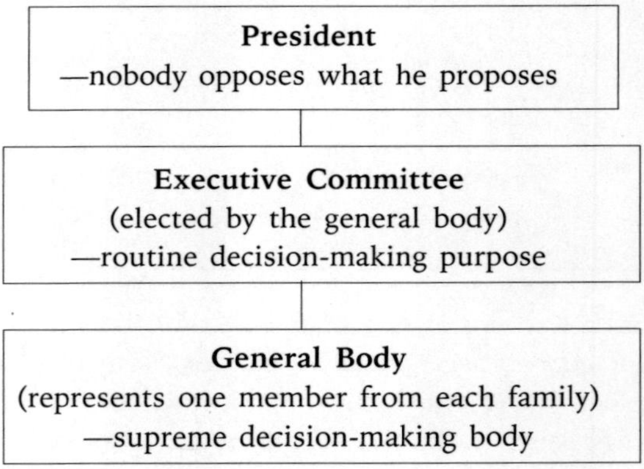

Analysis of People's Institution (PIs) in Forestry

Protection: The PIs evolved responding to the crisis of deforestation, which made the supply of forest resources very difficult. They did face challenges in their initial stages, but they have succeeded which has brought about a degree of confidence to their institutions. Protection by far is the most important initial activity and the JFM policy too lays stress on this. What was an open access resource is now being converted to a CPR by the PI (or co-operative) and naturally there is opposition to this from the free-riders. With the protection activity getting stabilised over a period of time, the difficulty of protection eased and the cooperative is seen to have a legitimate right over the resource they protect within a certain boundary. JFM confers legitimacy to the cooperatives and ensures that the PIs have a univocal right to protection of a defined resource.

Decision-making: Three types of rule are involved in the process of decision-making: a) constitutional, b) collective choice, and c) operational choice. The constitutional rule is that the degraded forestland would be protected together and managed as a CPR by forming PI, having membership of the residents of the village and a structure of executive committee nominated by the general body, which would then frame the rules for the governance of the CPR. The collective choices are determined at the general body and cover aspects such as protection, imposing of fines in case of violation, basis of distribution of benefits, appointment of the executive committee and so on. There remain great variations in the collective choices and rules among the different PIs. The complexity of collective choices and rules increases with the increase of the activities of the PI. The operative rules are specific to each PI and cover the mode of extraction of resources, timing and specific procedures to be adopted and so on. The number of operative rules also increases with the activities increasing and diversifying over time.

Activities: The activities of the PIs are seen to grow with the building up of confidence among the members. With the success of the protection activity, the forest starts regenerating and there is an increase in supply of resources. They begin some management activities and with its success their confidence increases. With some initial success the name of the PIs spread and some government departments approached them to take up other activities. Depending on the formation and the support available the PIs expand their activity profile.

Evolution and growth of PIs in forestry: The evolution and growth of PIs is analogous to a stone falling in still water. A set of ripples is immediately created. Similarly when a PI evolves it also creates a series of ripples in the development scenario. Its evolution may be catalysed by an NGO or Forest Department. PIs start with protection and with growing confidence takes on other activities. With this, the interface with the outside agencies grows and also becomes stronger. Each activity entails widening of contacts with the outside world and gaining recognition and credibility.

How the Areas Under Forest Management can be done more Efficiently

The existing method in a JFM model is that the group will be given 25 per cent of the share of the trees that will be fallen at the end of the maturity period of the trees in a specified area. This does not give any immediate returns to the poor forest dwellers. This does not actually give any incentive to these poor people to protect the forest. The model should be such in which the people get a regular source of income by being a member of the JFM committee. This method can be improvised by several steps:

1. A method should be devised in which each member will be given a particular area under him. The number of trees in that area will be made known to him. Now maintenance of the area will be evaluated by the Forest Department at the end of each 6 months. Proper maintenance will ensure him again 50 per cent of trees. This method will ensure regular source of inflow of money to the tribal people or any other forest dwellers. Again the area which he will be able to bring under reforestation scheme will again ensure him sole right on 50 per cent of the trees and rest 50 per cent will be under Forest Department, which they will sell and give as per norm—25 per cent of the booty—to the concerned person.
2. Giving news of the illegal wood traders and poachers each time will again ensure him some increase in the percentage of his share.
3. The forest dwellers will be taught the technique to raise particular crops in the rows of the forest, which will again help them to get a regular source of income. In the newly forested areas this method can be adopted intensively for continuous three to four years, so that it becomes a huge boost to reforestation as this will ensure them a double income source.
4. The members of the JFM committee will come under special public distribution system (PDS) where they will be given necessary items under special discount rate.
5. Some cheap discount store will be set up in those areas for the members of the JFM committee.
6. Free education for their children and health check-up scheme facility will be given to the members.
7. Entertainment like showing movies will be taken up once a week, where during interval they will be given education regarding importance of maintenance of forest trees highlighting the benefit they will get from the forests.
8. Regular campaigning (twice a month) involving the members against illegal wood traders and poachers will be done.
9. Most often it is found that there are some huge barren lands under Forest Department, these lands will be entrusted to the members of

the JFM committee and they will be given special training on how to reutilise the land.

10. Eco-tourism can be developed in the forest degraded areas.

References

Acheson, James M. (1993). *The Lobster Gangs of Maine*. Hanover: The University Press of New England.

Agarwal, A. and S. Narain (1989). *Towards Green Villages*. New Delhi: Centre for Science and Environment.

Axelrod, R. (1984). *The Evolution of Cooperation*. New York: Basic Books.

Baland, Jean-Marie and Jean Philippe Platteau (2000). *Halting Degradation of Natural Resources: Is there a Role for Rural Communities?* Rome: Oxford University Press.

Bandyopadhyay, J. and V. Shiva (1987). *Ecological Audit of Eucalyptus Cultivation*. New Delhi: Sagar Publishers.

Basu, K. and A. Mishra (1993). "Sustainable Development and the Commons Problem: A Simple Approach", in P. Bardhan, M. Datta-Chaudhuri and T. N. Krishnan (eds.), *Development and Change: Essays in Honour of K.N. Raj*. Bombay: Oxford University Press.

Binswagner, H. (1989). "Brazilian Policies that Encourage Deforestation in the Amazon", *World Bank Environment Department Paper* No. 16.

Brara, R. (1987). *Shifting Sands: A Study of Rights in Common Pastures*. Jaipur: Institute of Development Studies.

Bromley, D.W. (1992). "The Commons, Common Property Resources and Environmental Policy", in *Environmental and Resource Economics* 2.

Bromley, D.W. and M.M. Cernea (1989). *The Management of Common Property Natural Resources: Some Conceptual and Operational Fallacies*. Washington DC: The World Bank.

Chambers, Robert (1987). "Sustainable Livelihoods, Environment and Development: Putting the Poor Rural People First", *Discussion Paper* 240. Sussex: IDS. December.

Chambers, Rober, N.C. Saxena and Tushaar Shah (1990). *To the Hands of the Poor: Water and Trees*. San Francisco: West View Press.

Chopra, K., G.K. Kadekodi and M.N. Murthy (1989). *Participatory Development: People and Common Property Resources*. New Delhi: Sage Publications.

Commoner, B. (1971). "The Environmental Costs of Economic Growth", in R. Dorfman and N. Dorfman (eds.), *Economics of the Environment: Selected Readings*. New York: W.W. Norton.

Cordell, J. and M.A. Mckean (1992). "Sea Tenure in Bahia, Brazil", in D.W. Bromley (ed.), *Making the Commons Work, Theory, Practice and Policy*. San Fransisco: ICS Press. pp. 183-207.

Dasgupta, P. (1982). *The Control of Resources*. Oxford: Blackwell.

———. (1993). *An Inquiry into Well Being and Destitution*. Oxford: Clarendon Press.

Dasgupta, P. and K.G. Maler (1997). "The Resource Basis of Production and Consumption", in P. Dasgupta and K.G. Maller (eds.), *The Environment and Emerging Development Issues*, vol. I. Oxford: Clarendon Press.

Demsetz, H. (1967). "Toward a Theory of Property Rights", in *American Economic Review* 62.

Elwin, V. (ed.) (1963). *A New Deal for Tribal India*. New Delhi: Publications Division, Ministry of Home Affairs.

Fisher, A.C. and M. Hanemann (1997). "Valuation of Tropical Forests", in P. Dasgupta and K.G. Maller (eds.), The *Environment and Emerging Development Issues*, vol. I. Oxford: Clarendon Press.

Government of India (1999). *Economic Survey 1998-99*. New Delhi.

———. (1998). *Statistical Pocket Book: India 1997*. New Delhi.

———. (1988). *National Forest Policy, 1988*. New Delhi : Ministry of Environment & Forests.

Government of West Bengal (1998). *Economic Review 1997-98*. Calcutta.

Guha, R. (1989). *The Unquiet Wood: Ecological Change and Peasant Resistance in the Himalaya*. New Delhi: Oxford University Press.

Gupta, A.K. (1987). "Why Poor People Do Not Co-operate? A Study of Traditional Forms of Co-operation with Implications for Modern Organisations", in G.C. Wanger (ed.), *Politics and Practices of Social Research*. London: George Allen and Unwin.

Haimendorf- von Fürer, C. (1995). "Social Change in a Himalayan Region" in J.S. Lall (ed.), *The Himalaya: Aspects of Change*.New Delhi: Oxford University Press. pp.69-97.

Hardin, G. (1968). "The Tragedy of Commons", *Science* 162: 1243-248.

———. (1978). "Political Requirements for Preserving our Common Heritage", in H.P. Bokaw (ed.), *Wildlife and America*. Washington DC: Council on Environmental Quality. *ng*.

Indian Environmental Society/Ford Foundation (1990). "Executive Summary of the Workshop on Sustainable Forestry", *Forest Management Partnerships: Regenerating India's Forests*. September 10-12. New Delhi: The Ford Foundation.

Iyenger, S. (1988). "Common Property Land Resources in Gujarat: Some Findings about Their Size Status and Use", *Working Paper* 18. Ahmedabad: The Gujarat Institute of Area Planning.

Jodha, N.S. (1986). "Common Property Resources and Rural Poor in Dry Regions of India", *Economic and Political Weekly* 21(27): 1169-181.

———. (1987). "A Case Study of Degradation of Common Property Resources in India", in P. Blaikie and H. Brookfield (eds.), *Land Degradation of Society*. London: Methuen.

———. (1991). "Rural Common Property Resources: Contributions and Crisis", *Economic and Political Weekly* 25(26).

Kant, Shashi (2000). "A Dynamic Approach to Forest Regimes in Developing Economies", *Ecological Economics* 32: 287-300.

Lise, Wietze (2000). "Factors Influencing People's Participation in Forest Management in India", *Ecological Economics* 34: 379-392.

Nadkarni, M.V. (1993). *Agricultural Policy in India: Context Issues and Instrument*. Mumbai: Department of Economics & Policy, Reserve Bank of India.

Nadkarni, M.V., K.N. Ninan and S.A. Pasha (1994). *The Economic and Financial Viability of Social Forestry Projects*, Ford Foundation.

Noronha, R. (1997). "Common Property Resource Management in Traditional Societies", in P. Dasgupta and K.G. Maller (eds.), The *Environment and Emerging Development Issues*, vol. I. Oxford: Clarendon Press.

Ostrom, E. (1999). "Coping with Tragedies of the Commons", *Annual Review of Political Science* 2: 493-535, June.

Pasha, S.A. (1992). "Common Property Resources and the Rural Poor—A Micro Level Analysis", in *Economics and Political Weekly* 27(46).

Rao, V.M. (1991). "Growth in the Context of Development: Case of Dryland Agriculture", *Economics and Political Weekly* 26(13), March 30.

———. (1994). *Agricultural Growth, Rural Poverty and Environmental Degradation in India*: Delhi: Oxford University Press.

Redclift, Michael (1992). *Sustainable Development: Exploring the Contradictions*. London: Routledge.

Ribbentrop, B. (1900). *Forestry in British India* (Reprint 1989). New Delhi: Indus Publishing Company.

Sarin, M. (1998). "Community Forest Management: Whose Participation?", in I. Guijt and M. Kaul Shah (eds.), *The Myth of Ccommunity: Gender Issues in Participatory Development*. London, UK: Intermediate Technology Publications. pp.121-130.

Singh, K. (1994). *Managing Common Pool Resources: Principles and Case Studies*: New Delhi: Oxford University Press.

———. (1995). "The Watershed Management Approach to Sustainability of Renewable Common Property Resources", *IRMA Research Paper* 14.

Singh, K. and S. Bhattacharjee (1991a). "Privatisation of Common Pool Resources of Land: A Case Study of West Bengal", *IRMA Case Study* 6.

———. (1991b). "The Bergram Majhipara Common Pool Fish Pond: A Case Study", *IRMA Case Study* 8.

Singh, K. and V. Ballabh (1989). "Afforestation of Village Common Lands: A Case Study of Aslali Village Woodlot in Gujarat", *IRMA Case Study* 4.

Somnathan, E. (1991). "Deforestation, Property Rights and Incentives in Central Himalaya", *Economic and Political Weekly* 26(4). January 26.

Tiwari, D.D. (1994). "Developing and Sustaining Non-Timber Forest Products: Policy Issues and Concerns with Special Reference to India", in *Journal of World Forest Resource Management* 7: 51-78.

Vyas, V.S. (1991). "Agrarian Structure, Environmental Concerns and Rural Poverty", *Elmhirst Memorial Lecture,* XXI International Conference of Agricultural Economists, Tokyo, Japan.

Vyas, V.S. and V. Ratna Reddy (1993). "An Assessment of Environmental Policies and Policy Implementation in India", *International IDEAL Workshop,* Phillipines. Also variability of output.

Wade, R. (1988). *Village Republics: Economic Conditions for Collective Action in South India*. Cambridge: Cambridge University Press.

World Bank (1992). *World Development Report 1992*. Washington DC: The World Bank.

2

Economic Growth, Human Development and Protection of Common Property

Cross-State Analysis of Forest Sector in India

SACCHIDANANDA MUKHERJEE and
DEBASHIS CHAKRABORTY

Introduction

The common property resources (CPR) in rural India includes village pastures, community forests, waste lands, common threshing grounds, waste dumping places, watershed drainages, village ponds, riverbeds etc. A considerable section of the rural poor in India, especially the lower income group, is heavily dependent on the CPRs in general and on forest resources in particular for livelihood, e.g.—sourcing fuel, animal grazing etc. It has been observed that the employment and income generated from CPR-related activities play a key role in reducing rural inequality (Jodha, 1986).

Though the importance of other types of CPRs in India cannot be undermined, forests stand out as a category from the list, both in terms of value of natural resource as well as means of livelihood. Annexure A-2.1 shows the forest cover in the Indian states. It is seen from the table that forest cover is quite considerable in northeastern states, Jammu and Kashmir, Chhattisgarh, Goa, Jharkhand, Orissa etc. On the other hand, the forest cover is quite low in Punjab, Haryana, Rajasthan, UP, Gujarat etc., which includes both high and middle income states.

The research question in the current analysis is to analyse the underlying factors that determine the performance of a state in protecting its forest resources. The Environmental Kuznets Curve (EKC) hypothesis, which focuses on the relationship between income level of a country/region and its environmental sustainability, needs to be mentioned here. According to EKC hypothesis, growing income level of a region might initially be associated with environmental degradation, but at a later stage the demand for better environment would come and consequent adoption of superior pollution abatement practices would follow. In order to understand the evolving nature of the response of the states, the analysis is conducted for two periods: Period (A) ranging over 1990-1996 and Period (B) covering 1997-2004.

In line with EKC, it could be argued that a couple of related factors might also influence environmental performance of a country/state. For instance, enhanced economic growth in a country may get translated into economic development and improve human development (HD) level. The HD augmentation in the current period may in turn strengthen the perception about environmental sustainability in the subsequent period. Therefore, enhancement of HD may also be considered as a positive factor for improving environmental performance.

The current analysis is arranged along the following lines. After a brief literature survey on the interrelationship between environment, economic growth and human development, the data and the methodology of the present analysis are described. The ranking results are reported next, followed by the regression analysis. Finally on the basis of the findings, the conclusions are drawn.[1]

Relationship between Environmental Quality and Economic Growth

The literature on the relationship between per capita income (PCI), or the per capita net state domestic product (PCNSDP) in case of States within a country, and environmental degradation generally attempts to verify the existence of an inverted U-shaped curve in the PCI *vs*. pollution plane (EKC hypothesis). The relationship implies that environmental quality (EQ) worsens up to a certain income level, but improves afterwards with prosperity, as countries/regions shift to cleaner production technologies or focus more on pollution abatement (Andreoni and Levinson, 2001). Studies based on both ambient concentration of pollutants (Selden and Song, 1994; Grossman and Krueger, 1995) or actual emissions of pollutants (de Bruyn *et al.*, 1998; Bruvoll and Medin, 2003) support the EKC hypothesis. Recent studies show that while local pollutants like sulphur dioxide (SO_2), suspended particulate matter (SPM), carbon monoxide (CO) etc. support EKC hypothesis; other pollutants exhibit either monotonicity or a N-shaped relationship (Dinda, 2004).

In the Indian context, Mukherjee and Kathuria (2006) analysed the EKC relationship for 14 major Indian states over 1990-2001 by considering 63 environmental variables, arranged under 8 broad environmental groups. The findings indicated a slanting S-shaped relationship between EQ and PCNSDP,

1. The present analysis draws from the work of Mukherjee and Chakraborty (2009).

suggesting that economic growth (EG) has occurred in Indian states mostly at the cost of EQ. A similar non-linearity was reported by Mukherjee and Chakraborty (2009) as well.

Relationship between Environment and Human Well-being

It is increasingly believed that environmental problems should no longer be viewed as the side effects of development process but rather be an integral part of it (Ginkel *et al.*, 2001; Mukherjee and Chakraborty, 2011), which has been reflected in the Target 9 of the United Nations' Millennium Development Goals (MDGs). Melnick *et al.* (2005) argued that achieving environmental sustainability requires careful balancing of HD activities while maintaining a stable environment that predictably and regularly provides resources and protects people from natural calamities. It is also held that socioeconomic inequality leads to environmental inequality, which may consequently affect the overall extent of EQ (Boyce, 2003). Comparison of the environmental degradation index (EDI) created by Jha and Murthy (2001) for 174 countries with their HDI supported the existence of an inverted N-shaped global EKC rather than an inverted U-shaped one. In the Indian context, Mukherjee and Chakraborty (2009) also observed that HD improvement is generally conducive for cleaner environment.

Methodology

Determining the Environmental Quality Index for Forest Sector (EQIF)

The EQIF for the states is postulated to be linearly dependent on a set of certain observable variables relating to depletion and degradation of forest resources and has been determined by adopting the HDI method. The idea is that all these variables, when combined, give a composite indicator for the forest sector in the states, unobservable otherwise. We assume X_{ij} to be the actual value of the ith indicator for j^{th} state of India with respect to forest. In line with the HDI method, the indicators are transformed into their standardised form, by which the adjusted values of X_{ij} (i.e., EX_{ij}'s) to be used for the analysis become:

$$EX_{ij} = (X_{ij} - X_i^*)/(X_i^{**} - X_i^*)$$

where, X_i^* and X_i^{**} are the minimum and maximum values for the ith indicator. Now, $EQIF_j$, i.e., the environmental quality index score for the jth state with respect to forest is arrived at by summing the EX_{ij}s over i by using the following formula:

$$EQIF = \frac{1}{n}\sum_{i=1}^{n} EX_{ij}$$

The obtained *EQIFs* measures the forest-related achievement of the states, i.e., the states with higher scores are characterised by better forest endowment. The obtained EQI_js lead to the ranks of the jth state, where states having higher $EQIF_j$ gets higher rank.

Determining the Human Development Index (HDI)

Following the principle of the National Human Development Report (NHDR) 2001 (Government of India, 2002) methodology, for calculation of the HDI, three variables are considered, namely—(i) inflation and inequality adjusted per capita consumption expenditure (X_1); (ii) composite indicator of educational attainment (X_2); and (iii) composite indicator on health attainment (X_3). Following the HDI method, the HDI score for the jth state is then given by:

$$HDI_j = \frac{1}{3}\sum_{i=1}^{3} X_i$$

where, X_i represents the normalised values of the three indicators selected for construction of the HDI score, obtained by using the following formula:

$$X_i = (X_{ij} - X_i^*)/(X_i^{**} - X_i^*)$$

where X_{ij} refers to attainment of the ith indicator by the jth state and X_i^{**} and X_i^* are scaling maximum and minimum values of the indicators respectively (i=1 to 3).

The monthly per capita consumption expenditure data is obtained from National Sample Survey Organisation (NSSO) for two periods (1993-94 and 1999-2000), adjusted for inequality using estimated *Gini* ratios, and further adjusted for inflation to bring them to 1983 prices by using deflators derived from State specific poverty line (Raju, undated).

The composite indicator on educational attainment (X_2) is arrived at by considering two variables, namely literacy rate for the age group of 7 years and above (e_1) and adjusted intensity of formal education (e_2). The idea is that literacy rate being an overall ratio alone may not indicate the actual scenario, and the dropout rate needs to be incorporated in the formula. The data on literacy rate is considered for two periods, namely—1991 and 2001. The adjusted intensity of formal education (IFE) data is used for two periods—1993 and 2002. The following weightage is assigned for the two variables for determining the composite indicator:

$$X_2 = [(e_1 \times 0.35) + (e_2 \times 0.65)]$$

The adjusted IFE is estimated as weighted average of the enrolled students from Class I to Class XII (where weights being 1 for Class I, 2 for Class II and so on) to the total enrollment in Class I to Class XII. This is adjusted by proportion of total enrollment to population in the age group 6-18 years (Raju, undated). According to the formula suppose E_i be the number of children (rural and urban combined) enrolled in i^{th} standard in 2002, i=1 for Class I to 12 for Class XII). Then weighted average of the enrolment (WAE) from Class I to Class XII is derived by:

$$WAE = \frac{\sum_{i=1}^{12} i \times E_i}{\sum_{i=1}^{12} i}$$

Hence, the total enrolment (TE) of children from Class I to Class XII in 2002 becomes:

$$TE = \sum_{i=1}^{12} E_i$$

Then, the IFE for children (rural and urban combined) in 2002 is obtained as:

$$\text{IFE} = \frac{\text{WAE}}{\text{TE}} \times 100$$

From the IFE, we can determine the adjusted intensity of formal education (AIFE) for children (rural and urban combined) in 2002 by using the following formula:

$$\text{AIFE} = \text{IFE} \times \frac{\text{TE}}{\text{P}_\text{c}}$$

Where P_C represents the population of children (rural and urban combined) in the age group of 6 to 18 years in 2001.

The composite indicator on health attainment (X_3) is arrived at by considering two variables, namely Life Expectancy (LE) at age one (h_1) and the reciprocal of infant mortality rate (IMR) as the second variable (h_2). For h_1, which measures the life expectancy at age 1 (rural and urban combined), the two data points considered for the two periods are 1990-1994 and 1998-2002 respectively. On the other hand, the IMR (per thousand) data is considered for two periods, namely—1992 and 2000. The following weightage is assigned for the two variables so as to determine the composite indicator used for calculation of the HDI:

$$X_3 = [(h_1 \times 0.65) + (h_2 \times 0.35)]$$

Data

In order to obtain state-level secondary information on forest resources from published government reports and other databases for both the time periods, i.e., Period (A) and Period (B), the sample is restricted only to 14 major Indian states, namely—Andhra Pradesh, Bihar, Gujarat, Haryana, Karnataka, Kerala, Madhya Pradesh, Maharashtra, Orissa, Punjab, Rajasthan, Tamil Nadu, Uttar Pradesh and West Bengal. The current study considers only those indicators with at least two observations, where one of these observations is located within the boundary of the two sample periods. The selected indicators have then been normalised using appropriate measures of size/scale of the states (geographical area, population and GSDP at current prices).[2] The variables considered in the current analysis include:

1. Change in forest cover (dense and open forest) as percentage of geographical area (in percentage points): 1995 to 1997 and 1999-2001.
2. Change in per capita forest cover (dense forest, open forest, mangrove, scrub) (in hectare): 1995 to 1997 and 1999 to 2001.
3. Change in recorded forest area as a percentage of total geographical area: 1997 to 1999 and 1999 to 2001.
4. Change in common property forest area[@] as percentage of total recorded forest area: 1997 to 1999 and 1999 to 2001.
5. Change in common property forest area[@] as a percentage of geographical area: 1997 to 1999 and 1999 to 2001.
6. Change in per capita availability of recorded forest area (person/ha): 1997 to 1999 and 1999 to 2001.
7. Change in per capita availability of common property forest area (in person/ha): 1997 to 1999 and 1999 to 2001.
8. Change in protected area (national park and sanctuary) as a percentage of total geographical area: 1997 to 1999 and 1999 to 2001.[3]

The data on forest-related variables is sourced from CSE (1999), CSO (2000, 2002), issues of *Forest Survey of India* (1997, 1999 and 2001), data

---

2. Two key concepts need to be explained here, namely—endowment effect and efficiency in natural resource management effect. The depletion and degradation of natural resources and occurrence of environmental pollution is chiefly concerned with environmental management. On the other hand, the initial endowments of natural resources (forests, land and water) are determined by geographical, climatic and ecological factors. Quite understandably, the former is comparatively more influenced by human activities. By calculating the change in the natural resource position with respect to a base year, it is possible to isolate these two effects. The current study focuses on the environmental management efficiency effect as well as the size effect of the states.

3. *Note*: @ - common property forest area = protected + unclassed forest area.

published by Ministry of Environment and Forests etc. The data on economic growth in obtained from EPW (2003). The education-related data is obtained from NCERT (2002). On the health front, IMR data is taken from Sample Registration System Bulletins and LE data is taken from Indiastat database.

Results

Results on Environment

The forest-related achievements of the states are noted in Table 2.1. It is observed that Punjab, Haryana and Uttar Pradesh had been the toppers during Period (A), while the laggards include Bihar, Andhra Pradesh and Madhya Pradesh. However, over the period under observation a major change has taken place. During Period (B), Kerala, Orissa and Tamil Nadu have emerged as the top performers, while Uttar Pradesh, Bihar and Madhya Pradesh are now at the bottom. The decline in the performance of all three toppers during Period (A), i.e., Punjab, Haryana and Uttar Pradesh in Period (B) can be attributed to the increase in the agriculture-related activities within their territories and urbanisation fuelled by the growing population pressure. On the other hand, relative stagnation of the Orissa's economy has helped it to gain a higher rank during Period (B).

Table 2.1

Forest Quality Scores and Ranks of the States (Period A and B)

States	*Period A*		*Period B*	
	Score	*Rank*	*Score*	*Rank*
Andhra Pradesh	0.506	(13)	0.781	(9)
Bihar	0.515	(12)	0.480	(13)
Gujarat	0.594	(6)	0.769	(10)
Haryana	0.671	(2)	0.790	(7)
Karnataka	0.573	(9)	0.807	(5)
Kerala	0.517	(11)	0.942	**(1)**
Madhya Pradesh	0.158	**(14)**	0.230	**(14)**
Maharashtra	0.581	(8)	0.731	(11)
Orissa	0.584	(7)	0.864	(2)
Punjab	0.840	**(1)**	0.789	(8)
Rajasthan	0.520	(10)	0.804	(6)
Tamil Nadu	0.612	(4)	0.842	(3)
Uttar Pradesh	0.620	(3)	0.507	(12)
West Bengal	0.611	(5)	0.825	(4)

Results on Human Development

Table 2.2 provides the HDI scores and rankings of the states in the three sub-categories and the composite index for two periods. It is observed that while for Period (A), Kerala, Punjab and Maharashtra were holding the top three positions; in Period (B), Haryana had replaced Maharashtra from the top three. Kerala consistently performs well in all three categories. On the other hand, Punjab's performance in terms of consumption and health is good, but the same cannot be said on their educational attainments. Bihar, Uttar Pradesh and Madhya Pradesh are placed at the bottom for most of the categories.

Regression Results

Testing the Existence of the Environmental Kuznets Curve (EKC)

For testing the EKC hypothesis with respect to forest-related achievement of the states, a multivariate OLS regression model is performed. Different variants of the models are estimated by assuming the existence of non-linearity between PCNSDP and EQ. Apart from PCNSDP; several other explanatory variables are introduced to capture the dynamic aspects of EQ. The regression results summarised in Table 2.3 shows that there exists a linear relationship between forest-related achievement of a state and its income. In other words, rising income helps a state to protect its natural resources in a better manner. Among the control variables, EQIF is found to be positively related to a share of tertiary sector in GSDP (TERSHARE) and negatively related to a level of urbanisation within the state (URB). The result implies that if the share of tertiary sector increases in the economy (i.e., the share of agriculture and manufacturing sector decreases), then the forest-related achievement of a state increases. On the other hand, growing level of urbanisation denotes increasing pressure on natural resources and hence leads to depletion of forest resources. The linear relationship between EQIF of a state and its PCNSDP is shown with help of Figure 2.1.

Testing the Relationship between HDI Score and Environmental Performance

For analysing the relationship between the HDI score of a state and its EQIF, an OLS regression model is estimated by assuming the presence of non-linearity, which is summarised in Table 2.3. The regression result confirms the non-linearity between HD and EQIF. Among the control variables, while share of workers in agriculture (AGRWRK) is positively related to EQIF, level of urbanisation (URB) is found to be negatively related to it like in the earlier case. It implies that if the share of agricultural workers increase, the EQIF of a state may go up. The relationship between EQIF of a state and its HDI is

Table 2.2

HDI Scores and Ranks of the States (Period A and B)

States	*Consumption*		*Health*		*Education*		*HDI Score*	
	Period A	*Period B*	*Period A*	*Period B*	*Period A*	*Period B*	*Period A*	*Period B*
	(1)	*(2)*	*(3)*	*(4)*	*(5)*	*(6)*	*(7)*	*(8)*
Andhra Pradesh	0.338 (8)	0.196 (10)	0.300 (8)	0.410 (7)	0.136 (11)	0.344 (11)	0.258 (9)	0.317 (9)
Bihar	0.000 **(14)**	0.025 (13)	0.125 (11)	0.143 (11)	0.000 **(14)**	0.000 **(14)**	0.042**(14)**	0.056 **(14)**
Gujarat	0.575 (5)	0.636 (4)	0.275 (9)	0.374 (9)	0.484 (4)	0.531 (4)	0.445 (6)	0.514 (6)
Haryana	0.610 (3)	0.792 (3)	0.499 (4)	0.614 (3)	0.366 (8)	0.497 (6)	0.492 (4)	0.635 (3)
Karnataka	0.295 (9)	0.402 (8)	0.466 (5)	0.481 (5)	0.371 (7)	0.478 (8)	0.377 (8)	0.454 (7)
Kerala	0.831 (2)	1.000 **(1)**	1.000 **(1)**	1.000 **(1)**	1.000 **(1)**	1.000 **(1)**	0.944 **(1)**	1.000 **(1)**
Madhya Pradesh	0.052 (12)	0.000 **(14)**	0.000 **(14)**	0.000 **(14)**	0.165 (10)	0.396 (9)	0.072 (12)	0.132 (13)
Maharashtra	0.459 (7)	0.490 (6)	0.549 (3)	0.570 (4)	0.541 (2)	0.710 (2)	0.516 (3)	0.590 (4)
Orissa	0.258 (11)	0.069 (11)	0.083 (12)	0.089 (13)	0.235 (9)	0.377 (10)	0.192 (10)	0.178 (11)
Punjab	1.000 **(1)**	0.907 (2)	0.765 (2)	0.837 (2)	0.414 (5)	0.505 (5)	0.726 (2)	0.750 (2)
Rajasthan	0.294 (10)	0.307 (9)	0.241 (10)	0.312 (10)	0.038 (13)	0.317 (12)	0.191 (11)	0.312 (10)
Tamil Nadu	0.489 (6)	0.583 (5)	0.366 (6)	0.454 (6)	0.517 (3)	0.658 (3)	0.457 (5)	0.565 (5)
Uttar Pradesh	0.039 (13)	0.054 (12)	0.050 (13)	0.134 (12)	0.073 (12)	0.238 (13)	0.054 (13)	0.142 (12)
West Bengal	0.583 (4)	0.441 (7)	0.358 (7)	0.383 (8)	0.378 (6)	0.486 (7)	0.440 (7)	0.437 (8)

Note: Figure in the parenthesis shows the rank.

shown with the help of Figures 2.1 (a) and (b). It is observed from the figures that with rise in HDI, the EQIF of a state increases, but at a decreasing rate.

Figure 2.1 (a)

Relationship between Forest Quality Score of a State and Economic Growth

PCNSDP *versus* Forest

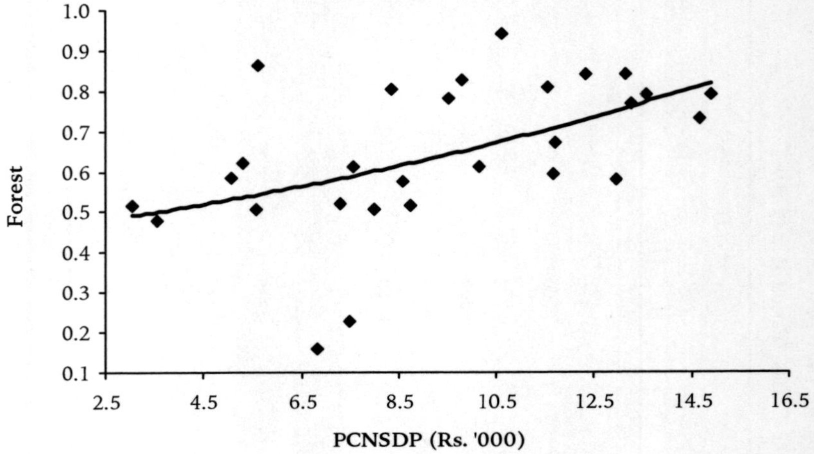

Figure 2.1 (b)

Relationship between Forest Quality Score of a State and Human Development

HDI Score *versus* Forest

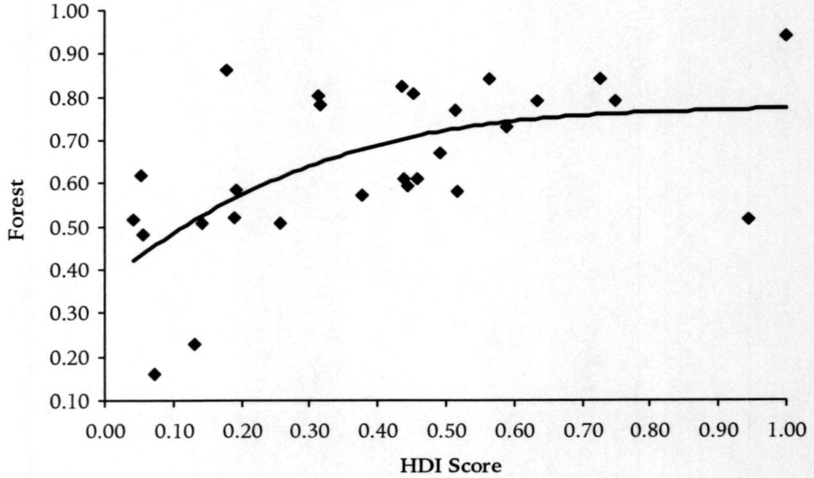

Table 2.3

Testing the Relationship between HDI, Economic Growth and Environmental Quality

Dependent Variable	*EQIF*		*EQIF*
Constant	-0.0640 (0.214)	1.0679 (0.2948) *	0.3321 (0.1214) *
PCNSDP	0.0724 (0.017) *	0.0715 (0.0205) *	
PCNSDP^2			
PCNSDP^3			
HDI			1.3688 (0.69)
HDI^2			-1.3020 (0.5882) *
HDI^3			
PRISHARE		-0.0116 (0.0054) *	
SECSHARE			
TERSHARE	0.0132 (0.0045) *		
URB	-0.0202 (0.0076) *	-0.0265 (0.0113) *	-0.0112 (0.0063)
AGRWRK			0.0097 (0.0056)
Adj. R^2	0.425	0.393	0.429
F-Stat	7.646	6.820	6.073
D-W Stat	1.557	1.590	1.906

Note: Figure in the parenthesis shows the White heteroskedasticity-consistent standard errors.
* implies coefficient is significant most at 0.10 level.

Observations

It is observed from the ranking of the states on their performance in protection of forest resources that a reorganisation of position is taking place over a period under consideration. The ranking of all three toppers during Period (A), i.e., Punjab, Haryana and Uttar Pradesh, declines in Period (B), which could be explained by the increase in agriculture-related activities within their territories and urbanisation fuelled by the growing population pressure. The notion is supported by the cross-state regression analysis on the relationship between forest-related environment and income. EKC hypothesis is rejected, as EQIF is found to be a linear function of income. However, the importance of the nature of growth on forest resources is underlined in no uncertain terms. For instance, it is observed that if the share of tertiary sector increases in a state (i.e., the share of agriculture and manufacturing sector decreases, leading to creation of less polluting activities), then the forest-related environmental quality of it improves. On the other hand, growing level of urbanisation creates increasing pressure on natural resources in a state and hence leads to the depletion of forest resources, thereby lowering its EQIF. The results indicate that economic growth needs to be associated with requisite level of environmental governance. Enhancing the participation of the local people in the conservation activities (e.g. joint forest management (JFM)) might provide a long-term solution in this regard. It could be noted from Annexure Table A-2.1 that in a number of states currently a considerable area is under the operation of JFM committees.

Looking at the relationship between human development and forest-related environment, it is observed that a non-linearity exists between the two, though the EKC type U-shaped curve does not emerge. Enhancement of human development is generally conducive for better forest-related performance for the states, which could be explained by the demand for better environment/ conservation with higher awareness. Therefore, the government initiatives for enhancing human development are likely to provide double dividend in terms of positive externalities through environmental benefits as well.

References

Andreoni, J. and A. Levinson (2001). "The Simple Analytics of the Environmental Kuznets Curve", *Journal of Public Economics* 80(1): 269–286.

Boyce, J.K. (2003). "Inequality and Environmental Protection", *Working Paper Series No.* 52. Amherst: Political Economy Research Institute, University of Massachusetts.

Bruvoll, A. and H. Medin (2003). "Factors Behind the Environmental Kuznets Curve: A Decomposition of the Changes in Air Pollution", *Environmental and Resource Economics* 24(1): 27–48.

Centre for Science and Environment (1999). "State of India's Environment: The Citizens' Fifth Report", (Part II: Statistical Database), New Delhi.

Central Statistical Organisation (CSO) (2002). *Compendium of Environmental Statistics 2002*. New Delhi: Ministry of Statistics and Programme Implementation, Government of India.

———. (2000). *Compendium of Environmental Statistics 2000*. New Delhi: Ministry of Statistics and Programme Implementation, Government of India.

de Bruyn, S.M., J.C.J.M. van den Berg and J.B. Opschoor (1998). "Economic Growth and Emissions: Reconsidering the Empirical Basis of Environmental Kuznets Curve", *Ecological Economics* 25(2): 161–175.

Dinda, S. (2004). "Environmental Kuznets Curve Hypothesis: A Survey", *Ecological Economics* 49(4): 431–455.

Economic and Political Weekly Research Foundation (EPWRF) (2003). *Domestic Product of State of India: 1960–1961 to 2000–2001*. Database software, Mumbai.

Ginkel, H.V., B. Barrett, J. Court and J. Velasquez (eds.) (2001). *Human Development and the Environment: Challenges for the United Nations in the New Millennium*. Tokyo, Japan: United Nations University Press.

Government of India. Forest Survey of India (1997, 1999 and 2001). "State of Forest Report", Dehradun: Ministry of Environment and Forest.

———. (2009). "Forest and Tree Resources in State and Union Territories", Dehradun.

———. (1999, 2001). "The State of Environment–India", New Delhi.

———. (2001). *India: National Population Policy. www.youth-policy.com*

Grossman, G.M. and A.B. Krueger (1995). "Economic Growth and the Environment", *Quarterly Journal of Economics* 110(2): 353–378.

Indiastat database website, accessible at *www.indiastat.com*

Jha, R. and K.V. Bhanu Murthy (2001). "An Inverse Global Environmental Kuznets Curve", *Departmental Working Papers: 2001–2002*. Canberra, Australia: Division of Economics, RSPAS, Australian National University.

Jodha, N.S. (1986). "Common Property Resources and the Rural Poor in Dry Regions of India", *Economic and Political Weekly* XXI (27), July 5.

Melnick, D., J. McNeely, Y.K. Navarro, G. Schmidt-Traub and R.R. Sears (2005). "Environment and Human Well-being: A Practical Strategy", UN Millennium Project, *Task Force on Environmental Sustainability*, available at: *http://www.unmillenniumproject.org/documents/Environment-complete-lowres.pdf* (last accessed on May 12, 2008).

Mukherjee, S. and D. Chakraborty (2009). "Environment, Human Development and Economic Growth: A Contemporary Analysis of Indian States", *International Journal of Global Environmental Issues* 9(1/2): 20–49.

Mukherjee, Sacchidananda and Debashis Chakraborty (2011). "Is Environmental Performance Influenced by Socio-Economic and Socio-Political Factors?: Cross-Country Empirical Evidences", *Sustainable Development* (forthcoming).

Mukherjee, S. and V. Kathuria (2006). "Is Economic Growth Sustainable? Environmental Quality of Indian States after 1991", *International Journal of Sustainable Development* 9(1): 38-60.

National Council of Education Research and Training (2002). *7th All India Educational Survey (AIES): All India School Education Survey (AISES)*, New Delhi: NCERT.

National Sample Survey Organisation (2001). "Differences in Level of Consumption among Socio-Economic Groups 1999-2000", NSS 55th Round, New Delhi: Ministry of Statistics and Programme Implementation.

———. (1997). "Differences in Level of Consumption among Socio-Economic Groups 1993-94", NSS 50th Round, New Delhi: Ministry of Statistics and Programme Implementation.

Office of the Registrar General (2001). "Census of India 2001", New Delhi: Ministry of Home Affairs, *CensusInfoIndia 2001* (Version 1.0)-database software.

Raju, S.D. (undated). "Measurement of Human Development – District Level", available at: *http://www.undp.org.in/hdrc/events/Aggregates/Goa/2* (last accessed on July 8, 2007).

Sample Registration System (SRS). *Bulletins*, New Delhi: Registrar General of India.

Selden, T.M. and D.S. Song (1994). "Environmental Quality and Development: Is there a Kuznets Curve for Air Pollution Emissions?", *Journal of Environmental Economics and Management* 27(2): 147–162.

Annexure Table A-2.1

Type of Forest Cover in Indian States

State	Forests as % of Reporting Area for Land Utilisation	Very Dense Forest	Mod. Dense Forest	Open Forest	Scrub	Reserved Forests	Protected Forests	Unclassed Forests	Area Under JFM Committees
	(%)	(sq. km)	(sq. km)	(sq. km)	(sq. km)	(sq. km)	(sq. km)	(sq. km)	(Ha)
Andhra Pradesh	22.59	820	24757	19525	10372	61210	1967	637	2290000
Arunachal Pradesh	93.74	20858	31556	14939	111	10546	9528	31466	90000
Assam	24.61	1461	11558	14673	179	17864		8968	80000
Bihar	6.65	231	3248	3325	134	693	5779	1	370000
Chhattisgarh	44.21	4162	35038	16670	107	25782	24036	9954	640000
Goa	34.63	511	624	1016	1	253	845	126	10000
Gujarat	9.91	376	5249	8995	1463	14122	479	4326	1700000
Haryana	2.61	27	463	1104	145	249	1158	152	56000
Himachal Pradesh	24.06	3224	6383	5061	327	1898	33060	2075	420000
Jammu and Kashmir	60.98	4298	8977	9411	2036	17643	2551	36	114182
Jharkhand	29.27	2590	9899	10405	683	4387	19185	33	2190000
Karnataka	16.10	1777	20181	14232	3176	28690	3931	5663	320000
Kerala	27.85	1443	9410	6471	58	11123	142		170000
Madhya Pradesh	54.31	6647	35007	36046	6401	61886	31098	1705	1100000
Maharashtra	19.27	8739	20834	21077	4157	49226	8195	4518	2500000
Meghalaya	27.23	701	5474	11105	1	1467	4171	11780	4000
Manipur	42.70	410	9501	7410	211	1113	12	8371	90000
Mizoram	77.10	134	6251	12855	1	7909	3568	5240	20000
Nagaland	28.17	1274	4897	7293	2	86	508	8628	20000
Orissa	37.33	7073	21394	20388	4852	26329	15525	16282	820000
Punjab	6.06	0	733	931	20	44	1153	1861	180000
Rajasthan	7.61	72	4450	11514	4347	12454	17416	2769	580000
Sikkim	37.32	500	2161	696	356	5452	389		10000
Tamil Nadu	16.43	2926	10216	10196	1206	19388	2183	1306	480000
Tripura	57.77	111	4770	3192	75	4175	2	2117	100000
Uttar Pradesh	6.92	1626	4563	8152	745	11660	1420	3503	80000
Uttarakhand	60.96	4762	14165	5568	271	24638	9882	131	860000
West Bengal	13.70	2987	4644	5363	29	7054	3772	1053	630000

Source: Forest and Tree Resources in State and Union Territories (2005) and State Forest Annual Report (2009).

3

Environment and Sustainable Development

New Perspectives, Challenges with a Vision for Future

ANJALI DEWAN

Introduction

Sustainable development is a socioecological process characterised by the fulfillment of human needs while maintaining the quality of the natural environment indefinitely. The linkage between environment and development was globally recognised in 1980, when the International Union for Conservation of Nature and Natural Resources (IUCN) published the *World Conservation Strategy* and used the term 'sustainable development'. The concept came into general usage following the publication of the 1987 report of the Brundtland Commission—formally, the World Commission on Environment and Development (WCED) set up by the United Nations General Assembly. This became the most often quoted definition of sustainable development as development that, 'meets the needs of the present generation without compromising the ability of future generations to meet their own needs.' It refers to the wise use of resources within a framework in which environmental, economic and social factors are integrated. It is about maintaining and improving the quality of life while safeguarding the quality of life of generations to come. The components of biological diversity have to be utilised in such a manner that it does not interfere with the natural functioning of ecological process and life-support systems. It is a collection of methods to create and sustain development which seeks to relieve poverty, create equitable standards of living, and satisfy the basic needs of people.

For India, a large country both populated and poor, to develop an environmentally sustainable development is not an option but a requirement. On one hand, India faces environmental degradation from poverty and population pressures, and on the other, it faces pollution from increased activities due to economic growth and the consequent changing consumption patterns. While the poor depend on the environment for their livelihood, the process of economic development relies on using natural resources to produce goods and services. The waste generated from consuming and producing these goods and services are in turn released back into the environment impacting it gravely. The environment provides security for present and future generations. The health of the environment is closely connected with the health of human beings, and it is economically beneficial for countries to prevent environmental degradation. The challenge therefore, in making development compatible with the environment is to restructure the economic system in a way that it will not destroy the environment even as economic progress continues.

Sustainable development does not focus solely on environmental issues. The 2005 World Summit Outcome Document, refer to the interdependent and mutually reinforcing pillars of sustainable development as economic development, social development, and environmental protection. Green development is generally differentiated from sustainable development. Green development prioritises what its proponents consider to be environmental sustainability over economic and cultural considerations. Proponents of sustainable development argue that it provides a context in which it improves overall sustainability where cutting edge green development is unattainable. For example, a cutting edge treatment plant with extremely high maintenance costs may not be sustainable in regions of the world with poor financial resources. An environmentally ideal plant that is shut down due to bankruptcy is obviously less sustainable than one that is maintainable by the indigeneous community, even if it is somewhat less effective from an environmental standpoint. Some research activities start from this definition to argue that the environment is a combination of nature and culture.

Figure 3.1

The Three Components of Sustainable Development

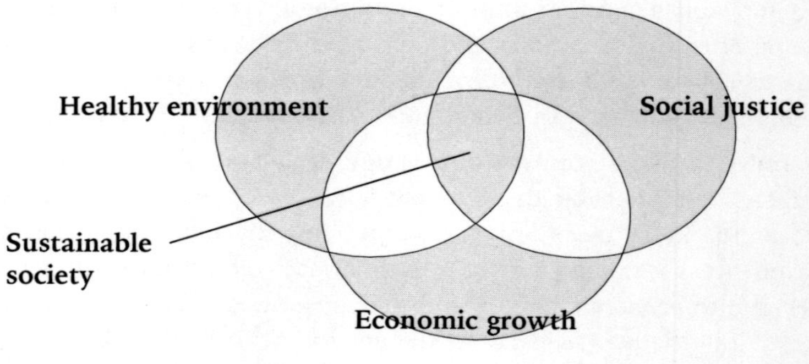

Environmental sustainability is defined as the ability of the environment to continue to function properly indefinitely. This involves meeting the present needs of human beings without endangering the welfare of future generations. The goal of environmental sustainability is to minimise environmental degradation, and to halt and reverse the processes they lead to. An unsustainable situation occurs when the sum total of nature's resources is used up faster than it can be replaced. Sustainability requires that human activity only uses nature's resources at a rate at which they can be replenished naturally. The long-term result of environmental degradation would be local environments that are no longer able to sustain human populations to any degree. Such degradation on a global scale could imply extinction of humanity.

Consumption of renewable resources	State of environment	Sustainability
More than nature's ability to replenish	Environmental degradation	Not sustainable
Equal to nature's ability to replenish	Environmental equilibrium	Steady-state sustainability
Less than nature's ability to replenish	Environmental renewal	Sustainable development

The gross mismanagement of our planet has much to do with an inequitable distribution of the benefits of development. Perpetuating this inequity can only mean a continuing drawdown on the world's natural resources and the environment. After a century of unprecedented growth, marked by scientific and technological triumphs that would have been unthinkable a century ago, there have never been so many poor, illiterate and unemployed people in the world, and their number is growing. Close to a billion people live in poverty and squalor, a situation that leaves them little choice but to go on undermining the conditions of life itself, the environment and the natural resource base.

The key socioeconomic pressures on the environment are population, poverty and illiteracy on one hand and economic growth, changing consumption patterns, structure of GDP and trends of urbanisation on the other. The state of India's environment with respect to air, water and land is well known. India is set to replace China as the most populous country in the world with 1.5 billion people by 2040. Since 1951 the population has grown from 361 million to 1 billion making every sixth person in the world Indian. As India has only 2.4 per cent of the world's total landmass, the population density is at 324 persons per square kilometre (Census, 2001) and pressures on the environment are correspondingly high.

Issues and Challenges

It is widely accepted that meeting the needs of the future depends on how well India and the rest of the world balances their social, economic and environmental goals when making decisions today. Research carried out under consecutive research framework programmes has addressed sustainable development, in particular, and environmental issues, in general, in a number of different contexts. The challenge therefore, in making development compatible with the environment is to restructure the economic system in a way that it will not destroy the environment as economic progress continues. Broadly speaking, the challenges facing us today are as follows:

- The ways in which greenhouse gas emissions and atmospheric pollutants affect climate, ozone depletion, soil and forests.
- The water cycle and soil-related aspects, especially looking at the impact of climate change on the hydro-cycle, interactions between land, ocean and atmosphere, distribution of groundwater and surface water, freshwater and wetland ecosystems, management strategies for water (availability, demand, vulnerability, quality, etc.).
- Understanding marine and terrestrial biodiversity to minimise the negative impact of human activities, relationships between society, the economy and biodiversity habitats, forecasting potential changes, studying the structures and mechanisms while assessing and managing the risks to biodiversity and ecosystems, as well as looking into conservation, mitigation and rehabilitation options for marine and terrestrial biodiversity.
- Mechanisms of natural disasters and desertification such as seismic and volcanic activity—and how these tie-in with climate change, land and soil degradation—and studying what science can do to help long-term forecasting of hydrogeological hazards, natural hazard monitoring, mapping and management, and to improve disaster preparedness and mitigation strategies.
- Strategies for sustainable land management, including integrated coastal zone management and assessment tools for measuring the impacts of agriculture and forestry on the environment, while looking at new technologies and integrated approaches to make these sectors more environmental-friendly.
- Systematic operational forecasting and modeling of climate change (atmospheric, terrestrial, marine-based) to improve understanding of extreme events, and to use data coming from the global monitoring for environment and security (GMES) programme and other global

observation systems for the climate, oceans and terrestrial habitat more effectively

- Cross-cutting issues, such as putting together sustainable development concepts and tools for measuring the sustainability dimension on relevant policies towards sustainable development in a very thorough, practical and meaningful manner.

Certain important factors which have an impact on the environment and sustainable development are as follows:

Socioeconomic Pressures on the Environment

There is pressure on the environment such as population, poverty, illiteracy on the one hand and economic growth, consumption patterns and urbanisation on the other.

Population

While the growth rate of the population has declined over the years, the population in India continues to be high on account of the large base of the population in the reproductive age group. The higher contribution to fertility due to the unmet need for contraception is estimated to be 20 per cent. High fertility rates are also seen due to high infant mortality rates, which are exacerbated by the lack of sanitation and health care facilities.

In addition, 50 per cent of girls marry below the age of 18, resulting in a typical reproductive pattern of 'too early, too frequent, too many' (GoI, 2001). The significance of population pressure is great as it impedes both development and environmental health.

Literacy

It is well known that literacy is important for every aspect of development and the environment is no exception. For instance, women's literacy reduces fertility rates and thus reduces population pressures. Literacy is essential for reducing poverty as it provides people with more opportunities to earn as well as increase their productivity. It also allows people to become more receptive towards alternative technologies and environmental management. People's ability to manage environmental resources like water, soil and forests improves with literacy. Thus, literacy helps reduce pollution as the added awareness allows people to link the health of the environment with their own health and well-being. In India, literacy rates are growing steadily but they are still far from adequate.

Poverty

Clearly, one of India's biggest problems is poverty. Fortunately, despite the growth of population and the low levels of economic development at the time of Independence in 1947, India has made significant progress in poverty reduction. The percentage of people below poverty line (BPL) has come down significantly. Yet, large numbers of people are still below the poverty line. The high incidence of poverty that still prevails in India underlines the need for rapid economic development to create more remunerative employment opportunities and to invest in social infrastructure of health and education.

The pollution that the poor generate is of a different kind than generated by the rich. The poor from lack of alternatives rely extensively on the environment for their needs; as a consequence its degradation affects their livelihood immensely. They, thus, become both victims and agents of environmental degradation as they both need and often exploit the environment to survive but suffer the most when it degrades. On the other hand, the poor are also victims of pollution. When water sources are polluted, the poor women have to walk further and spend considerable time fetching it. Lack of resources like fuels and safe drinking water results in more workload for the people. They also face health problems due to water borne diseases. Suffer from respiratory and eye diseases and their reproductive health is also affected. The rural population spends 15 billion workdays collecting water and fuels every year with serious health consequences. This should be seen as a tremendous national loss. How can India achieve 8 per cent economic growth on a sustainable basis if nearly 35 per cent of the population undergoes hardship and drudgery doing daily chores!

Urbanisation and Transportation

As countries industrialise, cities grow and traffic increases. This may reflect rapid economic development and industrialisation but also results in high levels of energy consumption and emissions. The progress of urbanisation is relatively slow in India compared to other developing countries and the bulk of the population continues to live in rural areas. The urban population increased from 19 per cent of the total population in 1965 to 28 per cent in 2000. It is expected that more than 50 per cent of population in India will reside in urban areas by 2025. The number of urban agglomerations/cities with populations of over a million has increased from 5 in 1951 to 9 in 1971 and 23 in 1991 to 35 in 2001. The increase in the requirements of urban transportation in India over a past few decades has outstripped the development of urban infrastructure. This has led to congestion of cities, proliferation of private vehicles, increased energy usage and increased pollution. The situation got worse due to shift of modal split away from railways towards road transport, and from public to private transport due to inadequate capacity expansion of public transport leading to rising emission levels of pollutants, and increased energy usage due to proliferation of private vehicles. Transport is a critical infrastructure for both intercity and urban development. When low-income groups do not have access to affordable transportation system, this imposes hardships and impacts their quality of life. Their time and energy is wasted in commuting, making them inefficient and thus sending them in a vicious circle of poverty and inefficiency. Efficient transport system is a critical infrastructure for the cities for economic productivity and quality of life. Transportation sector is responsible for a large share of air pollutants ranging from respirable particulate matter, nitrogen oxides, carbon monoxide, ozone and greenhouse gases etc. Therefore, in order to develop sustainability, greater emphasis needs to be placed on developing good public transport systems.

Water

Water pollution is a major cause of concern in India because in addition to causing ecosystem damage it adversely affects health and thereby impairs economic productivity of people. The current condition of water resources requires urgent action to be taken. For instance, 90 per cent of India's surface water resources are polluted to the extent that they are not fit for bathing. Also, about 200 million people do not have access to safe drinking water and utilisable water per capita is decreasing. The socioeconomic costs of water pollution are extremely high: 1.5 million children under 5 years age die each year due to water related diseases, 200 million peoples' days of work are lost each year and the country loses about Rs.3,60,000 million each year due to water related diseases.

Agriculture Industry Domestic Pollution

The three major contributors towards water pollution are the domestic, industrial and the agricultural sectors. Seventy-five per cent of the effluents by volume are from the domestic sector. This is because only 20 per cent and 2 per cent of waste water in Class I and Class II cities respectively is treated. Meanwhile, only 3.15 per cent of rural population has access to sanitation services and 115 million homes have no access to toilets of any type. In the industrial sector only 59 per cent of large and medium industries had adequate effluent treatment in 1995. In the agricultural sector, fertiliser use increased from 7.7 mt in 1984 to 13.4 mt in 1996 and pesticide use increased from 24 TT in 1971 to 85 TT in 1995.

One of the reasons why environmental standards are ignored is because they are seen to be expensive. However, when one calculates the cost of lowering pollution one must compare it to the health and economic benefits from abating pollution. For example, Rs.460 billion is needed to construct toilets in 115 million homes, wastewater treatment in 3696 cities/towns would cost Rs. 180-600 billion depending on technology, pollution abatement in industries would cost Rs.140 billion (about 1.2 per cent of total annual turnover). However, the loss from human health damages due to sanitation and water pollution is Rs. 360 billion per year. (Parikh *et al.*, 1996).

Air

Contrary to popular belief, air pollution is not an urban problem alone. While in cities suspended particulate matter (SPM), sulphur dioxide (SO_2) and nitrogen oxide (NO) levels are much higher than permissible limits, in rural areas indoor pollution kills half a million prematurely every year. Although in urban areas the transport sector is the highest contributor to air pollution, refuse burning and Industrial pollutants are also big contributors. It can be observed that SPM matter and respirable suspended particulate matter are at very high levels in both residential and industrial areas. In rural areas, air pollution is primarily produced indoor from the use of fire wood and other unclean sources of cooking fuel. Since 72 per cent of India lives in rural areas, this form of air pollution has significant impact on the population as a whole. Indoor pollution causes 0.41 to 0.57 million premature deaths per year.

The Need for a New Development Paradigm

The need for a new development paradigm was widely recognised by the mid-1980s. By then, the United Nations had already declared the second (1971) and third (1981) of its development decades a failure, especially in their inability to halt the cycle of poverty that existed in the world's poorest and slowest developing countries. Further, the conditioned loan policies of the World Bank and the International Monetary Fund (IMF) in combination with patterns of development assistance provided through various bilateral and other arrangements succeeded in plunging already impoverished countries into even deeper debt. Thus, the combination of high debt, slow economic growth, and rapidly increasing poverty enhanced the profound levels of human suffering that already existed in many of these countries. Groundbreaking events, such as the United Nations-backed Rio Earth Summit on Environment and Development, in 1992, and the Johannesburg Summit on Sustainable Development a decade later in 2002, show that policymakers recognised the grave challenges that lie ahead in preserving the planet's fragile environment. One of today's environmental challenges is to find ways of strengthening the scientific and socioeconomic perspectives to help authorities make decisions and produce sustainable development strategies.

Key Documents

The Rio Declaration: A statement of 27 principles upon which nations agreed to base their actions in dealing with environmental and development issues. The Rio Declaration built on the previous Declaration of the United Nations Conference on the Human Environment which was adopted in Stockholm in 1972. The Stockholm conference was the first global environmental meeting of governments, which stated that long-term economic progress needs to be linked with environmental protection.

Agenda 21: An action blueprint on specific issues relating to sustainable development that emerged from the Rio Summit. Agenda 21 explained that population; consumption and technology were the primary driving forces of environmental change and for the first time, at an international level, explicitly linked the need for development and poverty eradication with progress towards sustainable development

The Brundtland Report: 'Our Common Future' which was produced in 1987 by the World Commission on Environment and Development, also laid the foundations for the environmental principles. This landmark document highlighted that people needed to change the way they lived and did business or face unacceptable levels of human suffering and environmental damage.

Global Compact: The environmental principles provide an entry point for business to address the key environmental challenges. In particular, the principles direct activity to areas such as research, innovation, cooperation, education, and self-regulation that can positively address the significant environmental degradation, and damage to the planet's life support systems, brought by human activity.

Sustainable Development Strategies

The Government has emphasised the need for a coordinated approach to sustainable development. Its efforts to advance sustainable development in India and around the world are also evidenced by its action on climate change, human health and the environment; its support for the building and maintenance of green infrastructure and its recognition of the unique relationship tribal people have with the natural environment. In addition, the Government has also recently made a commitment to regularly update strategies, which will ensure steady progress. Some of the common themes that have emerged in this regard include:

- Enhancing the capacity of Indians to act and make decisions in support of sustainable development; strengthening and improving access to knowledge in support of integrated decision-making.

- Raising the level of understanding/awareness of sustainable development issues; providing the necessary tools to policymakers to a better advanced sustainable development.
- Building on existing partnerships and creating new ones; providing leadership in areas such as greening of operations; fostering innovation and technology and promoting the adoption of sustainable development policies and practices in India and abroad.
- A coordinated approach, which outlines best practices for departments to adopt in procurement, waste management, water conservation, energy efficiency, vehicle fleet, land use and human resource management needs to be prepared.
- Eco-efficient business practices that conserve energy, eliminate use of toxic materials, and reduce or reuse resources, innovative eco-efficient practices, tools, technologies and products can increase productivity while improving environmental performance.

The quality of life of Indians is intrinsically linked to the health and well-being of their communities. A sustainable community is adaptable to change and has a vision for the future. However, to become sustainable, communities face complex environmental, economic and social challenges and require a coordinated approach to information, support and expertise.

Sustainable Development and Healthy Indians

Human health is dependent on a wide range of factors including the natural and built environments, and social, cultural and economic conditions. Certain populations like children, pregnant women and lactating mothers, the elderly and tribal people may be particularly vulnerable. A meeting with representatives from a range of sectors needs to be initiated to discuss the implications of human health when making decisions for sustainable development. The intention is to continue this dialogue whereby departments will explore information needs, strategies for action and approaches to engage the public in order to protect the health and overall quality of life of our people.

Social and Cultural Aspects of Sustainable Development

There is growing recognition that the concept of sustainable development is multifaceted with links to social and cultural aspects, such as poverty, distribution of wealth, diversity, equity, gender and health. These not only may affect the environment, but also influence and shape economic behaviour, consumption patterns, and the longer term viability and sustainability of our society. We are at an early stage of understanding the nature and implications of the social and cultural dimensions of sustainable development. Accordingly, various governmental and private agencies will be working together to understand this area and better the relationships involved by undertaking exploratory research and providing social policy perspectives.

Environmental Quality Management Strategies

Environmentally sustainable development requires a combination of six strategies: (i) economic incentives against polluting, (ii) law enforcement, and technological interventions such as (iii) cleaner technologies, (iv) institutional mechanisms, (v) poverty alleviation programmes, and (vi) people's participation. This section examines how these strategies can be effective.

Law Enforcement and Other Controls

Consolidate pollution control laws: The environment has to be addressed holistically as different media like air, water and soil are interconnected and pollution once generated, can shift in space and time. That is, pollution can be diluted in more water or air or converted into different forms of pollutants like burning solid waste and causing air pollution or pollution laws must be consolidated to address this issue.

Introduce full liability laws: Industries and other polluters should be made fully liable for their pollution. After the Bhopal gas tragedy this should have been obvious, yet the Bhopal victims have not got adequate compensation.

Make clean technologies mandatory in new industries: The current industry growth rates are around 7-8 per cent. In this light, industries being established now will comprise 50 per cent of the market in 9 years (Jayaraman, 2001). Cleaner technologies should be made mandatory in such industries so as to reduce pollution in the coming years. Industries are less willing to change their techniques once they are established. Hence, they should be made to take on cleaner technologies right from the start. The technology for small scale industries can be improved by requiring large scale capital goods manufacturers to produce less polluting equipments.

Make functioning treatment facilities mandatory: Often, due to mandatory requirements, effluent treatment plants are set up but are not operated to save operating costs. Therefore, laws should provide for monitoring operation of these facilities.

Require environmental audits for industry: Environmental impact assessments are often made at the start of the project but prescribed environmental management practices are often forgotten once the project is underway. Regular audits would act as self-monitoring and enhance compliance to standards.

Provide effective right to information: If people have the information about what their neighbouring industry pollutes, they would generate pressures for abatement and treatment. Full liability laws need to be complemented with right to information.

Role of Pollution Boards

Central and state pollution control boards need to be strengthened technically—by installing modern equipments and training, financially—with larger budgets to hire better staff and facilities to monitor the pollution, and politically—by keeping them independent and free from political influence. This has to be done at all levels e.g., at municipalities and *gram panchayat* levels to control pollution. Suggestions have also been made that industries should be roped into monitor pollution and assist new industries in choosing environmentally-friendly technologies. Involving private sector in water supply and sewage treatment in urban areas may prove to be beneficial.

Role of Centre for Environment Education (CEE)

The primary objective of CEE is to improve public awareness and understanding of the environment with a view to promote the conservation and sustainable use of nature and natural resources, leading to a better environment and a better quality of life. It develops innovative programmes and educational material, builds capacity in the field of education for sustainable development. It undertakes demonstration projects in education, communication and development that endorses attitudes, strategies and is committed to ensure that due recognition is given to the role of education in the promotion of sustainable development. The activities and programmes of CEE have been guided by certain strategies for maximisation of quality, effectiveness and impact. These are as follows:

- Adaptability to different geographical, cultural, social and economic contexts.
- Partnerships utilising complementary strengths of other organisations to avoid duplication of effort and the network effectively for synergistic convergence of ideas and goals.
- Building synergies among government, NGO's and CEE for comprehensive impact.
- Identifying key entry points for different thrust areas and key targets for initiating and consolidating gains to achieve a multiplier effect.
- Facilitating networks to local, national and regional levels through a number of tools such as dialogues, directories, newsletters etc.

- Working to develop a cadre of professionals in order to improve and strengthen professionalism in the field of education for sustainable development (ESD) by capacity building individuals who in turn would infuse this professionalism into the organisations where they work to the mutual advantage of both those organisations and CEE.
- Ensuring quality control and excellence in the production of all material through in-house infrastructural support like the studios, workshops, printing facilities.

International Aspects of Sustainable Development

Ten years ago at the Rio Earth Summit, more than 100 states proclaimed 'sustainable development' to be fundamental to social and economic progress for all nations. The Indian government is addressing key sustainable development issues in the international arena, chief among them are: globalisation and competitiveness, environmental concerns such as persistent organic pollutants, regional conflicts and their impact on the environment, and the trade and environment interface. Our country is also working with developing countries to help improve their sustainability. Finding ways to improve the capacity of developing countries to negotiate and implement multilateral trade and environmental agreements is of particular interest at this time. CEE has developed international linkages with numerous agencies, both governmental and non-governmental, that are working in related fields. The linkages are in the form of collaborative projects, exchange programmes, training, symposia and seminars. CEE has international offices in Australia and Sri Lanka.

Fourth International Conference on Environmental Education

The first intergovernmental Conference on Environmental Education (EE) was held under the aegis of UNESCO-UNEP in 1977 at Tbilisi, USSR. Conferences have been held every decade thereafter at Moscow in 1987 and Thessaloniki in 1997. The fourth International Conference on Environmental Education (ICEE), co-sponsored by UNESCO and UNEP, was organised by the Government of India. CEE was the nodal agency for the conference which was held in the month of November at Ahmedabad. Being held during the United Nations Decade of Education for Sustainable Development (DESD 2005-2014), the conference looked at building a sustainable future. The recognition that education is a critical agent of transformation in changing lifestyles, attitudes and behaviour, increasing participation, visioning and realising a sustainable world and facilitating the use of communication, education, participation and awareness (CEPA) to foster the changing needs in different sectors needs to be further strengthened. Reflection, future vision, sharing are the crucial elements of ESD.

Economic Incentives

Appropriate pricing: Natural resources are often sold at a very low price, leading to their exploitation. For example, the subsidies on irrigation water have led to planting of highly water-intensive crops in regions inappropriate for this kind of agriculture. Excessive use of water has also resulted in water logging as well as depletion of ground water table making the soil saline. Removing inappropriate subsidies is essential to maintaining natural resources and would encourage development of more environmentally-friendly alternatives. People have also begun to accept appropriate prices for natural resources. Pricing water to reflect its scarcity will encourage users to use it more sparingly.

Tax based on pollution load: Presently, effluent standards are based on best available technology for specific industries. Industries have no incentive to improve standards in such a system. Instead, a pollution tax should be levied so that industries pay taxes in proportion to the pollution they generate. Such a policy will reduce pollution at source and can only work if there is effective monitoring and punishment.

Higher credit rating for green industries: Higher credit rating for green industries will encourage upcoming industries which are dependent on the market for capital to be more environmentally conscious. Once right to information and liability laws become effective, rating agencies would take care of this.

Reduce subsidies on fertilisers and pesticides: The current subsidies on fertilisers and pesticides do not ensure that they are used sparingly. Recently, pesticides and fertilisers were found even in bottled soft drinks indicating that the runoff from agriculture is contaminating groundwater at very high levels. Reducing fertiliser subsidies will encourage more controlled use.

Technological Interventions

Technological intervention for environmental management does not necessarily imply new inventions. Many environmentally-friendly interventions are traditional methods or simple techniques, which have been known but not used.

- Efficient irrigation: Since 84 per cent of all water in India is used for agriculture, efficient irrigation is the best method to deal with water wastage. For example, applying water to the roots of crops through drip irrigation saves a considerable amount of water, fretless, pesticides, and electricity for irrigation. It also prevents soil erosion or water logging.

- Integrated pest management: Using integrated pest management (targeting the insects using natural methods) instead of pesticides would reduce pollution greatly. Vermiculture has been shown to be an effective method to deal with organic solid waste, which is becoming a major problem in urban areas. If the community can be made to sort their garbage, this can also provide organic manure.

The Path Forward

As the sustainable development journey continues, the various departments will continue to learn and improve. There will be further efforts to coordinate between departments as we move to the implementation of the updated strategies. This will include efforts related to communication and awareness building amongst our countrymen, science and technology development and staff training. The Government will also move forward on the current priority areas and identify opportunities for additional collaboration. The transition to a more sustainable society in the 21st Century will not be achieved by one department or one level of government alone. Moving towards more sustainable development will require the commitment and attention of decision-makers at all levels of society. The sustainable development strategies are an important element of the GoI's response to this significant challenge. The strategies will identify common goals and actions towards ensuring healthy human and natural environments, supporting emerging governance systems, advancing appropriate use of natural resources, building vibrant communities, and influencing international activities. Sustainable development is about a more inclusive society, which provides for better protection of the environment and use of natural resources and shares the benefits of economic growth as widely as possible. It is about achieving a better quality of life.

It is not possible for the government to monitor pollution and the corresponding acts of all industries and individuals. People must be made stakeholders in the environment through awareness campaigns. Industries are sensitive to public pressure. Experience in the West suggests that firms wish to maintain a green profile when citizens are aware of environmental issues. Through generating awareness, the public could directly affect the environmental practices of industries. India has to move forward technologically in an overall context to avoid pollution at source. Modern technologies already developed in the developed countries such as better power plants and cleaner vehicles should be considered through technology transfer for pollution measurements. Technological innovations allow India to develop rapidly in an environmentally-sustainable manner. India cannot afford to neglect the environment, if sustainable development is desired. The pressures on environment have to be curtailed by reducing population pressure, increasing literacy, environmental awareness drives and poverty alleviation programmes. The poor have to be turned into agents for environmental restoration by involving them in forest management, waste management, recycling and so on in a manner that creates incentives for them to use natural resources in sustainable manner. The economic activities must be conducted using environment conserving and resource saving technologies. Managing environment through better urban designs, improvement in transportation infrastructures and creative use of information technologies needs to be considered seriously. The strategy for environmental governance should consist of law enforcement, providing economic incentives, people's participation, institutional reforms and support, and technological improvements. With the determination of the government, private sector, NGOs, and people, India can perhaps achieve sustainable development.

References

Alger, C.F. (1990). "Grass-roots Perspectives on Global Policies for Development", *Journal of Peace Research* 27(2): 155-168.

Brandt Commission (1980). *North-South: A Programme for Survival*. London: Pan Books.

Burek, D. (ed.) (1991). *Encyclopedia of Associations: International Organisations*. New York: Gale Research Inc.

Campfens, H. (1990). "Issues in Organising Impoverished Women in Latin America", *Social Development Issues* 13(1): 20-43.

Census of India (1993). "Emerging Trends of Urbanisation in India: An Analysis of 1991". New Delhi: Govt. of India.

Central Water Commission (CWC) (1998). *Water Statistics of India*. Delhi.

Census of India (2001), Series-1 *India*, Paper-1 of 2001.

Economic Survey of Delhi (1999-2000).

Ester, R.J. (1993). "Towards Sustainable Development: From Theory to Praxis", *Social Development Issues* 15(3): 1-29.

Jayaraman, K.S. (2001). "Indian Regulatory System Stifles Industry Growth", *Business and Regulatory News* 19(2): 105-106.

———. (1995). "Mexico City's Water Supply: Improving the Outlook for Sustainability". DC, Washington: National Academic Press.

Midgley, J. (1990). "International Social Work: Learning from the Third World", *Social Work* 35(4): 295-301.

Nayak, R.K. and H.Y. Siddiqui (eds.) (1989). *Social Work and Social Development*. New Delhi: Geetanjali Publishing House.

Parikh K.S., M. Panda and N.S. Murthy (1996). "Modelling Framework for Sustainable Development: Integrating Environmental Considerations in Development Strategies", *Discussion Paper* No. 127. Mumbai: IGIDR.

World Commission on Environment and Development (1987). *Our Common Future*. New York: Oxford University Press.

World Bank (1990). *World Development Report 1990: Global Poverty*. New York: Oxford University Press.

———. (1993). *World Development Report 1993: Investing in Health*. New York: Oxford University Press.

II

Managing Common Property: Case Studies

4

Institutional Issues in Self-Initiated Forest Management Systems of Jharkhand in Eastern India

BHASKAR MITRA

Background

Forest policies in India have undergone several important shifts in paradigm over the past 125 years that has gradually reoriented the national sector strategy towards first centralisation and later devolution of authority for the management of the forest lands. From the early days of 'Scientific Forestry'[1] introduced by the British, the shifts in forest policies have clearly been from a centralised bureaucratic system to that of a decentralised one. The nationalisation of much of India's forest land in the latter part of the 19th century that had initiated a process of alienation of all those forest dependent communities from the management of designated public lands, who had for generations lived in harmony with it and continued enjoying its usufruct rights, till the resolution on joint forest management (JFM) was accepted by the Ministry of Environment and Forests (MoEF), Government of India in 1990.[2]

JFM policies and programmes, the direct outcome of the 1988 National Forest Policy, were designed to encourage communities to participate in the protection and management of the public forest domains. JFM programmes were established by means of the Government of India resolution of June 1990, after the Central government and various states passed enabling legislations that endorsed the protection responsibilities and usufruct rights of villages over small tracts of public forestlands. The programmes attempted to bring government support to these small, informal community organisations by providing them with greater legitimacy. Forest department (FD) strategies

1. A term coined essentially by the German and British foresters to exemplify the silvicultural techniques developed in forest management.
2. The National Forest Policy of 1988 paved the way for the JFM order of 1990.

included supportive actions such as formulating new policies, inventorying and registering forest protection committees (FPCs), village forest committees (VFCs) and eco-development committees (EDCs),[3] demarcating boundaries, assisting village groups to frame micromanagement plans, and running extension programmes. By June 2005 the MoEF estimated that 99,868 JFMCs were controlling access to more than 21.44 million hectares of degraded natural forest (MoEF and Winrock, 2005). The JFM programme has received large financial support from external funding agencies. Nearly Rs. 49 billion has been received for forestry projects since the 1990s in various states. The major funding agencies have been the World Bank, OECF-Japan, DFID-UK, SIDA-Sweden, EEC, UNDP and Germany.

At present, approximately 23 per cent of India's land area or 76 million hectares are officially designated forestlands. The area under actual forest cover, however, is estimated at approximately 67 million hectares (FSI, 2003), more than 90 per cent of which are under the administration of the State Forest Departments (Khare *et al.* 2000; Mukherjee, 1995).

While this entire process of policy change at the national level was taking place, several self-initiated forest protection groups[4] (SIFPGs or one form of community forestry groups) displayed how community forest protection took place at the grassroots. Community forest protection is fundamentally a decentralised grassroot level movement initiated largely by small villages (and in some case hamlets) to protect local natural forests from further degradation. It is the control over forest uses by a more or less well defined group of people claiming customary use rights (Ascher, 1995). Common features of all community forestry initiatives are that they support the control, management and use of forest and tree resources by local communities; they respect the social, economic and cultural relationships between people and forests; and they involve a decentralised and participatory approach to forest management, which assumes that the best stewards of the world's forests are the communities living in and around them (FAO, 2000). Protection activities are thereby co-ordinated through traditional and informal cultural institutions, as opposed to more centralised government mechanisms.

The SIFPGs that form the institutional setup of the self-initiated forest protections systems (SIFPSs), which operate in many states like Gujarat, Jharkhand, Orissa and Karnataka, have demonstrated the efficiency of this system, which adopts effective forest conservation practices. SIFPSs have evolved independently, without any assistance from the government, as a result of acute shortage of biomass and the states' inability to fulfill the basic fuel

3. Together they are known as the JFMCs.

4. Coined for the first time in Poffenberger *et al.*, 1996.

wood and fodder requirements of the affected communities. These institutions have stood the test of time and have successfully managed in staying together, regenerating their degraded forest and at the same time drawing economic benefits out of it. In this given situation, it is necessary to assess and learn from the SIFPSs that have evolved over the years, for their effective replication in other areas, as well as incorporating the learning points into the JFM programme (Ravindranath *et al.*, 2000).

For several years, the SIFPGs did not have any policy-level support from the government. Following the first JFM order in 1988, there were amendment guidelines for strengthening JFM in the years 2000 and 2002. The guidelines to strengthen the JFM order of February 2000 provided some breathing space for the SIFPGs. To integrate the SIFPGs into the JFM programme through the recognition of the groups, the order stated that:

> The community groups in many places in Orissa, Bihar, Gujarat, Andhra Pradesh and Karnataka are performing the essential functions of forest protection and regeneration. These groups need to be identified, recognised and registered as JFM committee after proper verification of records and enquiry. The period of their existence and duties performed for protection and regeneration of forests should be assessed and proper weightage given to them for deriving benefits under the JFM programme.

Before this, there had been no attempt to integrate the SIFPGs in the mainstream forestry institutions. With this landmark decision, it was hoped that it would act as a breather for the SIFPGs operations in various parts of the country as also help them in getting funds for forest management activities being carried out by them.

Objectives of the Study

The larger study (of which this paper is an outcome) hypothesises that community forestry under self-initiated forest protection systems (CFSIFPSs) have been successful wherever communities have had a strong institutional base and realised the economic and ecological perils of forest degradation. The study also hypothesises that a very strong institution at the grassroots leads to the creation of informal sanctions favouring forest conservation thereby resulting directly in the regeneration of the forests and indirectly in the economic upliftment of the community. The study adopts a systemic approach wherein one looks at the institutional, ecological and economic aspects of forest management systems. This paper looks at issues that may have lead to the success of the self-initiated forest protection institutions.

The Study Location and Methodology

The area where the fieldwork was carried out is the state of Jharkhand in India, one of the most forested regions of the country and with a very rich heritage of tribal culture. The two field sites are situated in Kudada and Bero in East Singhbhum and Ranchi districts, in the state of Jharkhand, respectively. The literal meaning of Jharkhand is 'land of forest'.

Geographically both the study sites were very similar. Besides being located in a similar socio-cultural-ecological setting, both consisted of a patch of forest in the centre protected by the communities residing in villages that formed some sort of a complex of villages situated within or along the periphery of the forest. In Kudada, there was a cluster of around 32 villages in the vicinity of the forest while at Bero, there were 20 villages located likewise. The villages that were taken up for the study are given below:

1. Harhanji-Khuratoli
2. Nagri
3. Baridih
4. Hariharpur-Jamtoli
5. Jahanabaj
6. Balidih
7. Kudada
8. Talsa
9. Turamdih
10. Dari

Thus, a total of 10 villages were covered during the study.

The systems approach used in this study required collection of three types of data relevant to the study—social and institutional, ecological and economic. The methodology for collection of data for each of the components was therefore different. A range of qualitative and quantitative data was collected using a mix of various methodologies.

Findings

In June 2000, the amendment to the JFM resolution mentioned that all SIFPGs that were operating in various parts of India needed to be brought under the JFM fold. Soon after the formation of Jharkhand in year 2000, the government of Jharkhand passed a new JFM legislation, necessitating the formation of new JFMCs all over the state. Towards meeting the above objectives, several steps were taken with limited success. Given the changing political and policy scenario, this section captures the impact it has had on the institutions at Kudada and Bero and on the overall sustainability of the forest. It also captures the dilemmas that engulf the community and also tries to understand the future course.

Confusion between Various Institutional Formats

Some community members of the SIFPGs at Kudada were unaware of the JFM[5] programme as the Forest Department has been approaching them for converting the SIFPGS into JFMCs.[6] The community members were visibly not too keen on being a part of the JFM programme and some members had vociferously resisted the formation of the JFMCs in the region. This had also antagonised the Forest Department and the relationship with the department had become sour. Nevertheless, the JFMCs were formed (at least on paper) in most of the study villages and the two institutions now exist side by side.

The JFMCs were formed by grouping contiguous villages or hamlets while under the SIFPGs, *sakha samitis*[7] were formed based on single or at most two hamlets, rather than on multiple hamlets/villages. In the present scenario, because villages have become units of developments, the JFMCs formed by the FD mostly consisted of 1, 2, or even 3 villages. The states' Forest Departments have not formed hamlet level committees anywhere in India including Jharkhand, as it increases the administrative burden. As a result, the fact that the JFMC members do not attend meetings is not surprising as executive committee members often belong to different villages. On the other hand, it is much easier for community members within a *sakha samiti* to communicate as they belong to the same hamlet and therefore it does not involve travelling for anyone.

The institutions at Bero were more strongly rooted than at Kudada, due to their linkages with the traditional *parha*[8] system. The *parha* system now seems to have been weakened due to political interference, this inspite of the fact that *panchayat* elections have not happened in Jharkhand. Although there have been efforts made by several individuals in the region to uphold the *parha* system, there has definitely been a weakening in this traditional village administrative mechanism. As forests were traditionally part of the villages, they were managed by decisions taken by the *parha* system, and every member

5. In Jharkhand there are two types of JFMCs—eco-development committees (EDCs) in areas within five kilometres of sanctuaries, national parks and reserved forests and forest protection committees (FPCs) for protected and other forests.

6. The Forest Department in Uttaranchal has also been complained of similar attempts at converting the *van panchayats* into JFMCs.

7. At Kudada one could see a two-tiered structure consisting of a central *kendriya samiti* and hamlet level *sakha samiti*.

8. The *parha* system is seen as a village-level administration system within the Oraon communities.

of a *parha* had to obey and follow the orders of the *parha raja*. But with the gradual weakening of the *parha* system, the traditional system of forest management is gradually loosing its foothold and it is being replaced with the highly politicised Panchayati Raj system.

Community members at both sites also expressed concerns about the manner in which the JFM committees were formed. There was a general perception that JFM created rules (drafted rather than crafted) that neglected existing and prudent uses of natural resources, local knowledge and cultural contexts. There was also lack of consultation with senior community members in the process of formation of JFM committees. In general, there existed a lot of suspicion in the minds of the people regarding the interests of the Forest Department.

Defined Forest Boundaries

The SIFPGs at Kudada and Bero operated within well-defined but largely undemarcated boundaries. The location of the villages where the SIFPGs have been operating is such that an overall clearer picture of the boundary is evident to the communities. At Kudada, the forest boundary was restricted to the visible range of the forest from the village and was thus not demarcated the way it is done in current practices within JFM. However at Bero, the communities have accepted the boundaries as chalked out by the Forest Department at different points in time.

Clear definition of boundaries operates upon the notion of exclusion principle, whereby those residing outside the sphere of the resource base have no right of appropriation of the resource. Exclusion principle that SIFPSs operate upon works on the principle that villages that are not part of the SIFPS are non-locals or outsiders. The exclusion principle at Bero and Kudada essentially operates with the assumption that locals would be able to restrict the influence of non-locals over access to the resource. However, it was evident that it had failed as the number of 'illegal' collectors of forest produce had only gone up. Therefore, an inclusion principle[9] would be a more correct or practicable solution where locals could also allow non-locals limited rights of appropriation as well as a say in the management of the resources.

Congruence between Making of Rules and Local Conditions

The institutions at Bero and Kudada interacted with the forest resources through a set of working rules that may be broadly termed as membership, decision-making and operational rules something which is very similar to the constitutional, collective and operational rules enunciated by FAO (2000).

9. The inclusion principle is a more egalitarian concept than the exclusion principle, which local communities in any case find very difficult to implement.

1. The membership (or constitutional) rules were rules determining membership in a collective decision-making unit. These set of rules determined the membership rules for creating, changing and revoking the rules of collective decision-making rules. At both the sites, it was seen that every household was a member of the SIFPGs by virtue of being members of the *gram sabha*.
2. The decision-making (or collective) rules were for making operational rules in which the SIFPGs and the JFMCs can make, modify or revoke operational rules. In both the cases it was seen that covertly it was the traditional administrative unit (*gram sabha* or its federating body) that were responsible for the collective rules. The JFM rules allowed for formation of communities responsible for framing constitutional rules. But what is important here is to understand how rules are made by collective decision-making. In case of the JFM programme, quorum must be complete for conducting meetings and passing of resolution. The decision-making process is simpler in case of the SIFPG wherein the *gram sabha* meets and decisions are taken through a simple majority. As was seen in both the sites, attendance in *gram sabha* meetings was a must and absence was taken as an acceptance of the resolutions being taken.
3. The operational rules were rules that guided behaviours concerning the forest resources. These rules defined as to who gets access to the resources, limits to harvest and contribution if any to be made by the community for the management of the resource. At both sites it was seen that all community members residing in the study villages had access to forests. The JFMCs and the SIFPGs allowed for free collection of fuel wood and non-wood forest produce (NWFP) for household use as well as specifying areas where they were to be collected from. Apart from the above, patrolling was a contribution in kind to be made by the communities in exchange for benefits from the resource, for ensuring the exclusion of non-locals from appropriation of the resource.

In the case of the SIFPGs, the rules framed were primarily of informal nature (unwritten), that is, they are not codified laws and regulations that are issued by the legislative process or formal decree. These rules have been derived from customs or practices. On the other hand, those put forward by the JFM programme are all codified and legislated. There is no reason to believe that such formal rules would be enforceable just because they are codified. Infact, rules were found to be more enforceable when backed by social norms.

At Kudada it was also found that formal rules (JFM rules) were in direct conflict with informal rules and therefore disobeyance of formal rules was higher. However when enquired about rules regarding forest management at Bero, the respondents listed down the JFM rules in the hope of providing more 'approved'[10] answers.

Questioning Feasibility of Exclusion and Scope of an 'Inclusion' Principle

At Bero there are problems related to both appropriation and provision of the resource while at Kudada it is primarily of appropriation nature. Appropriation problem deals with the production relationship between yield and the level of input which is assumed to be given. The problem is to allocate the yield. It is applicable to the case of the fuel wood and the NTFPs that are flows. On the other hand, provision problems deal with the assignment of responsibility for building, restoring and maintaining the resource system over time and the focus is mainly on the stock aspect of the resource system, in this case it is the forest. At the same time, it needs to be understood that fuel wood and NTFPs are substractable resources i.e., when one harvests it; it is substracted from the total body of the resources available. Effective rules are required to allocate scarce and substractable flows productively. Pricing could be an option that could be adopted at Bero and Kudada, however at times it could also be counterproductive leading to some individuals grabbing as much of the flows as possible and resulting in non-economic use of the service and highest level of conflict among users. Thus in Bero and Kudada, the resource involved being highly subtractive, institutional arrangements being related to flow allocations are intimately linked to sustainability of the resource base. This relates to the exclusion principle i.e., preventing non-residents from appropriating the resources.

The feasibility of exclusion however has an important impact on people's incentives to care for the resource. In general the more feasible it is to limit access, the more the community feels that they have tenure security i.e., people would gain from the benefits of the property. A higher feasibility of exclusion is associated with stronger incentives to nurture, protect and invest (money and time) in a particular resource. Given the location and characteristics of the resource base at both the sites, exclusion of non-locals (those residing in the immediate vicinity of the target villages) over harvest of the forest produce is a highly contestable issue. Though an exclusion principle barring non-locals from collecting products from the forest is in place, yet it was not very strictly adhered to. The problem of controlling access

10. But even otherwise, since the JFM programme was in a better state of acceptance at Bero and had been in implementation here for a longer period of time, the community at Bero was more aware and approving of the programme.

has been very difficult in the case of forest patches that were on the periphery of the villages. This coupled with the moral economy[11] attached to the issue of exclusion, has made feasibility of exclusion low. The feasibility of exclusion however varied from resource to resource and between seasons. For example there were fewer collectors of NWFP from the forest as compared to the fuel wood and exclusion was stronger between November to May (the agriculture lean season), when communities could carry out patrolling vigorously. Communities at Bero and Kudada have tried the exclusion principle to prevent non-locals from collecting harvestable from their forests. Exclusion is cheap but very difficult to implement under conditions of resource scarcity, especially if the non-locals who move in do not have any resource base of their own to fall back upon. Institutions at Kudada (and not Bero) have been able to control the free-riding problem by assuring their members that—benefits they could receive would exceed the cost, collective benefits could be achieved only through member contribution in terms of time, and all beneficiaries contribute their share of needed inputs. But the challenge of introducing an inclusion principle now seems imminent.

Ever since community forestry went underway in the 1990s, there has been debate on what constitutes 'local community' and subsequently to whom management authority was to be shifted. This essentially is putting 'exclusion principle' in different words. Primarily it is therefore a debate between 'local villages' and 'non-local villages' over the management of the resource. Grahn (2008) states that an inclusion principle would encourage a more rational decision-making for sustainable utilisation of the fragile ecosystem. Her study in Mali supports developing an inclusion principle between sedentary (with excess of resource base) and migrant users of resources (with shortage of resource base). Strong influences of an 'inclusion principle' were also seen during the initial JFM experiments in Arabari, West Bengal (Mukherjee, 1995). Thus empirical evidence points to a strong case for an inclusion principle whereby an effort may be made to tie a bond between 'local communities' and 'outsiders'. However the definition of 'outsiders' must be clearly defined within limited boundaries for this theory to be successful and there is a need for a larger debate on this issue.

Nature of Consumption

Forests are characterised by renewability but with a relatively long gestation period. If the time horizon for sustainability is distinct and not visible, then appropriate rules may be the difference between creating a sustainable conjunctive use system and destruction of the resource base. The

---

11. Most of the non-locals are in some way related to these locals. Feasibility of exclusion therefore gets coloured under relationship and moral terms.

nature of consumption like the feasibility of exclusion creates incentives and disincentives that influences the way people behave towards resources. Infact in Bero due to the influence of outsider (non-locals) threats over the forest, a large number of locals have also played a role in the unsustainable extraction of fuel wood from forest for sale in the Bero market (Personal Communication, Simon Oraon 2003).

Over the years the marginal cost of extraction of forest produce for rural communities (especially women) in terms of drudgery has gone up as a result of the open access treatment of the forests (Sarin *et al.*, 2001). At Bero especially, in the face of non-local extraction of forest resources, the locals have also started over-harvesting the resource base due to the fear of loosing it all out to the non-locals. Thus it can lead to a potentially lose-lose situation for both locals as well as outsiders. The situation needs to be transformed into one of win-win situation otherwise sustainability of the resource would be threatened. As Feeny (1988a) puts it '...without having alternative choices or some form of local autonomy over available resources, rural people have every incentive to cut down the trees before someone else would do so.' This also calls for additional gearing up on part of the Forest Department, which is currently completely inadequately motivated and trained to carry out this job.

Monitoring

Monitoring both in case of the SIFPGs at Kudada and Bero was through regular patrolling carried out intensively during the non-agricultural seasons. The mode of patrolling at some of the places was regular activity, while at a few other places it was more of an informal nature. Usually it was seen that patrolling was an activity carried out in conjunction with the forest produce collection. However women at Kudada complained that due to strict patrolling carried out, they had difficulties in collecting fuel wood for household use. Patrolling was however weak in some villages at Bero where the JFMCs were not very strict in implementing the same. Inspite of monitoring and subsequent action on illegal fuel wood collection there was an element of philosophy attached to the whole activity that can be explained only through concepts of moral economy (Thompson, 1993).

Apart from the philosophical angle attached to the issue, there was also an element of fear involved in preventing outsiders who often threatened the locals with dire consequences. Male patrolling members complained that women offenders were more difficult to deal with than the male offenders. Patrolling teams have often been charged with sexual offences by women head loaders leading to clashes between communities. Therefore patrolling as an activity has been through a series of periodic highs and lows.

Graduated Sanctions

The JFM programme allows for graduated sanctions on insiders indulging in illegal collection and trade in fuel wood. Even the SIFPGs at Kudada have claimed that often forest offences were carried out by insiders against whom graduated sanctions in the form of social or financial penalties were slapped. The primary form of sanctions resorted to by the communities were fines, treating community members to meals or in extreme cases ostracism from the community. At both sites the communities also claimed that dealing with insiders was relatively easier than dealing with the outsiders (also see Rasmussen and Dick, 1995).

However the issue of outsiders was a trickier one to handle. The Indian Forest Act of 1927 states that the right of compounding in case of forest offences lay in the hands off the Forest Department. The JFM programme states that collection of headload for consumption from forest was free, but there is little mention of the role to be played by the local communities if outsiders collected fuel wood from the forest. Villagers claimed that fuel wood collection is often carried out on an unsustainable scale whereby the green branches are cut and left to dry for 3-4 days and collected later under the garb of collecting dead, dying and diseased parts. Thus implementing sanctions on offending villagers (both locals and non-locals) has been difficult.

Group Size and Collective Action

Volumes of literature on collective action have discussed the negative association between groups' size and collective action (Agrawal, 1998; Potetee and Ostrom, 2004). Yet Kudada has been able to deal with the constant growth in the size of the organisation by creating levels of sub-groups (*sakha samitis*) thereby reducing the increased complexity of task and the difficulty of coordination that is brought on by large membership. Kudada could however face problems in the future due to the fact that large groups often churn out new leaders who may enter into conflicts with one another. The other threat is the fact that the *sakha samitis* have been situated at large distances, and over such large area of geographical operation, anti-group behaviour often goes unnoticed and unpunished. Such a situation is less of a probability at Bero, where the area of operation of the *parha* is limited to specific number of villages (like 12 or 21 villages etc.) through cultural affiliations; however, Bero is affected by a problem of heterogeneity that afflicts the region.

Chamberlin (1974), Oliver and Marvell (1988), Pecorino (1999), Gautam (2002) and Agrawal (1994) do mention that till a point, increase in group size could improve coordination. They demonstrated that level of provision of normal collective good increases with group size, despite decrease in the number of individual contribution. Under these conditions, the number of individual contribution required to provide the collective good may decrease with the size of the group. However, as Buchanan and Tullock (1962) puts it that with very large groups probably the cost of coordination would outweigh the benefits from increase in size. Thus when discussion involves formation of federations out of primary groups, the same logic of group size applies.

If a group is smaller, and other things being equal, it should be less costly for members of the group to recognise each other and hence it becomes easier for the group to detect rule infraction by group members and also entry to the commons by no group users. The cost of decision-making and coordination of activities should similarly be related to group size. In such situations four factors have been identified: (i) cost of intragroup enforcement, (ii) cost of extra group exclusion, (iii) cost of decision-making, and (iv) cost of coordination. The cost involved for each of these activities is affected by more than group size and in particular response to the cost of transportation and communication. A situation similar to this was observed in the case of conflicts taking place at Kudada among the different members of the *kendriya samiti*. At Bero, on the other hand, the perceived dependence on forests has also led to lower levels of participation in its management by a particular section of the rural elite.

For efficient group action, the perception of marginal cost (costs for protection, patrolling duties etc.) must be less than the marginal benefits (benefits from forests). However the system breaks down when some members (including the elite) stop paying the marginal costs, leading to a higher input being made by the remaining individuals. However if everyone shares the marginal benefit (including those who did not bear the marginal cost) of the produce, that is when the system collapses.

Heterogeneity and Collective Action

Social heterogeneity is attributed to reduced chances of successful and continued collective action. However there is enough evidence to suggest that on the contrary heterogeneity could lead to stronger institutions through the emergence of strong leadership. Heterogeneity may be looked at in five different forms: (i) heterogeneity of endowment, (ii) political heterogeneity, (iii) wealth and entitlement, (iv) cultural heterogeneity, and (v) economic heterogeneity (Pearson and Laitin, 1996; Ghate, 2002). Scholars of micro-institutional solutions to common problems have long argued the capacities of local communities in crafting institutional arrangements. There is also recognition of the fact that local solutions are difficult to achieve when demographic changes are quick, local community is not dependent on the resource in question, substantial heterogeneity of interest exists, has little local autonomy to make and enforce rules exist and resource systems is large, all of which seems to be very relevant to both the study sites.

In the sociocultural backdrop of India and especially states like Jharkhand, cultural heterogeneity is often the biggest deterrent to collective action. At Kudada, there is no obvious reason for any form of cultural heterogeneity, as people in this region mostly belong to the tribal groups and therefore show great deal of unity. However there still remain some areas of concern especially when the various tribal groups are represented unequally at the legislative levels, who then wield their powers to create wedge between the power factions among the different tribal groups in the villages.

Bero on the other hand has to operate with a greater cultural diversity amidst a very active political environment (Ranchi district where it is located, is the capital of the state). At Bero, the traditional institutions gave weightage to the JFM programme, and moulded themselves as per the given norms. However once the traditional institutions were replaced, the mode of functioning of the Forest Department allowed for dominant groups (both socially and economically)[12] and their allies to take over the leadership of the JFMCs, relegating the traditional leadership to the background (Kant, 2000). The caste factor played an important role in Bero in opening up of the JFMCs and allowing the dominant groups (not necessarily numerically, but economically well-off and vocal groups), who had little concern and interest in forest protection work, to take over. Thus the old institutions were obliviated while the new institutions are still trying to establish themselves.

Kudada and Bero amply reveal that with differing levels of heterogeneity, collective action may be higher for groups where traditional administrative mechanisms are been adhered. Although heterogeneity is a real time threat, these local groups have tried to devise institutions that enable them to draw on complementarities to build stronger foundation for collective action as also throw up new leadership. When mechanisms break down, often due to political interference as is the case at Bero and Kudada, collective action breaks down as well.

Conflict Resolution Mechanisms

Throughout the colonial period, conflicts between the community and the Forest Department were a regular affair. There are several reports of conflicts between the communities and the Forest Departments in parts of Kumaon and the Madras presidency, during the British period. Popular resistance to state forestry embraced the forms of unrest ranging from confrontation with the authority, such as breach of forest law, as well as organised rebellion that

12. The Sadans about whom references have been made earlier.

challenged the right of the state to own and manage forest areas (Gadgil and Guha, 1992). However institutions also suffer from conflicts due to the various stakeholders groups that make up social milieu of the village, each with their own set of requirements from the forests. Current development thinking accepts conflicts as a reality and is considered often a good outlet for sorting out local differences. It must be mentioned here that conflicts often lead to greater accountability and may be a good social process (Pendzich *et al.*, 1994), but if not resolved in time may lead to ultimate breakdown of the systems (Bruce, 1998).

Internal conflict resolution mechanism within the locals was extremely efficient at both sites. At Bero the conflicts are initially resolved at the level of the JFMCs/SIFPGs and the *gram sabha* within the villages. If conflicts occurred between villages of the same *parha*, then *parha* meetings were called to resolve the issues. It was only when conflicts occurred between villages of different *parha*s the situation would get complex. It may be mentioned here that the systems of *parha* has been gradually eroded due to the implementation of the Panchayati Raj system.[13] Thus, it would be inadequate to blame only the Forest Department or its JFM programme for the breakdown of the SIFP system at Bero, but the Panchayati Raj system is to be equally blamed.

At Kudada too, conflicts were initially resolved at the level of the *sakha samitis*. If such conflicts went unresolved, then they were taken to the *kendriya samiti*. Such conflict resolution mechanisms are very low-cost, low-intensity mechanisms and have the potential to arrive at acceptable solutions in a timely and effective manner. However conflict-resolution mechanism at both the sites were complicated due to politicisation of issues, whereby political parties often backed offenders.

However the issues of concern were the areas where the communities entered into direct conflict with the outside groups like the one at Kudada where the conflict was between two parties of vastly differing strengths. Uranium Corporation of India Ltd. (UCIL) has in the past few years taken over large chunks of land in villages of Turamdih and Talsa. Village Marchagora, where the SIFPG movement initially began, and village Banduhurang has been completely annihilated by UCIL. More than half of village Turamdih now comprises of project affected persons (PAPs), who have been asked to relocate so that mining work could resume. The SIFPG is trying hard to prevent the further incursion of UCIL into the forest at Kudada. It is essential here that the advocacy and negotiations skills of the community members need to be worked upon to arrive at any amicable settlement of the issue.

---

13. The *panchayat* operates on the basis of population size, wherein a population of approximately 10,000 comprises a *panchayat*, unlike the *parha* system that operates on the basis of identical cultural identity.

But in the various conflicts that erupt over access and ownership issues of a resource, the Forest Department has an extended role to play. Without the active support of the Forest Department, the community would not be able to operate on its own. Adams *et al.* (2002) state that in the context of such misunderstandings there is need for dialogue between the various groups of stakeholders, as each of them views the problem in their own way, as is seen with the case of the UCIL taking over the village land for mining purposes. For the SIFPG at Kudada, the battle against the UCIL so far has been a losing one.

Process of Upscaling by Federating or Decentralising

There are two ways in which an institution like the SIFPGs spread and form nested enterprises—either through federating several smaller groups or through a process of decentralisation.[14] At both Kudada and Bero there are clear-cut evidences of decentralisation wherein the SIFPGs were first formed at one place and thereafter neighbouring villages were motivated to form their own committees. This effort of decentralisation is completely in tune with current development thinking on governance. One of the most important lessons from Kudada and Bero has been multi-tiered institutional arrangement that has emerged at the grassroots. Similar efforts have been carried out by several NGOs in Orissa where they have promoted the formation of FPCs and subsequent federated bodies. Briksh o Jeevan Bandhu Parishad (BOJBP) in Orissa is one such initiative towards the formation of federations (a process of building up i.e., several small institutions uniting to form one large institution) that now covers several systemic organisations and individual committees. Similar experiences are also heard from case studies in Gujarat in western India and the case of Van Samakhya in Andhra Pradesh.

The SIFPGs are part of larger institutions that has its embeddedness in traditional forms of multi-tiered village administration, thus leading to a vertical linkage. But for optimal efficiency, institutions must also connect laterally to other institutions that have stakeholders of the same order. In this regard, the NGOs and the Forest Department could be a good vehicle for such lateral linkages leading to the overall growth and sustainability of the institutions. The community at Kudada also incorporates the continuous process of adjustment across these nested level of rules at all these three levels of decision-making.

The institution building process could be better followed-up through the registration of such bodies under appropriate laws. Such efforts have been made in the case of Federation of Community Forestry Users (FECOFUN) in Nepal and Van Samakhya in Andhra Pradesh. Such registration would provide

14. One is a process of building up and the other is a process of building down, and both are equally effective.

these bodies with the much-needed legitimacy to function more efficiently. There have been opinions that possibly the best results could be achieved by registering such bodies under the Societies Registration Act of 1860 (GoI, 1860), which enables them to function as voluntary organisations that could at a later stage also access funds for the development of the region through submitting proposals to donor agencies. However framing of new laws for registering such bodies, whose capacities are yet limited, is a must that would strengthen their capacities for the roles of forest development.

'Limits to growth' is a principle that must be remembered as size may be a limiting factor to the process of upscaling. Studies by Agrawal (1994) confirm that very large groups are seldom sustainable as visibility reduces and leads to increase in conflicts within the group. There must be greater debate on discussing the pros and cons of upscaling, before it is implemented on large scale.

Conclusion

Globally forestry has evolved from that of centralised forest regimes to that of partial or complete decentralisation. Experiences from countries like Indonesia, Nepal, Thailand, Mexico and Chile show that much of the developments in the forestry sector in the past two decades have taken place in synchronisation with current development paradigms of participatory development or people centred development. Forestry would sooner or later have taken the same route was only too evident. What is interesting is the fact that the process has unfolded itself in a diverse manner. In some places, community rights over resources were merely acknowledged while at other places rights were actually handed over to them in reality.

But the acknowledgement of the community rights is far from over, especially in countries like India where positive steps have been taken but the devolution is far from over. The case of India is most complicated due to its own myriad of problems like population pressure, multiple stakeholders each with their own rights and fanciful ways of seeing things and regional and national politics. As a result, the forest dependent communities are the worst losers, as they are the ones living with the resource, while ways and means of managing the same are done elsewhere, by someone else on their behalf!

The communities situated at Bero and Kudada have been managing the forests for decades, while the policymakers take key decisions over the well-being of the resources, without taking into cognisance the well-being of either the resource base or the communities that are dependent on it. As it currently stands, the situation is explosive with the real threat of the forests being completely denuded in both the sites within the next two decades. The threat to Kudada is much more due to rapid land acquisition by UCIL mines. Having accepted the fact that the traditional custodians of forest must be given their due share, it must be accepted that it is easier said than done. Like communities elsewhere, it needs to be understood if at all these communities, who have been managing the forest in the past, would be able to do so in the future. Where is the guarantee that common property resources would not turn into open access resources, for appropriation by all and sundry! This study throws up some key lessons learnt from the field:

- Accept traditional territorial regimes.
- Accept traditional institutions and acknowledge local leadership.
- Strengthen traditional institutions through their capacity development.
- Promote traditional forest management rules.
- Success at hamlet-level institutions.
- Allow for spread of traditional institutions and promote networking.

Forest management in Jharkhand took root much before formal and organised forest management systems were initiated anywhere in India. When the reservation of forest began during the colonial period, it was through carving out of forestlands, most of which was open access lands, although not specified under designated community use. In some areas, it was conceded that village communities had right to collect forest produce and these rights were recorded and the area designated as protected forest. Tribal groups in Jharkhand like the Mundas, Ho's, Oraons, Santhals, who comprise a substantial chunk of the population, were managing their own forests under various arrangements of forest management since long. Such traditional or SIFPSs as referred to by many are still existent in many parts of Jharkhand even today. They also throw up interesting lessons for the incorporation within the next generation JFM programmes in India:

- Allow for flexibility in institutional mechanisms.
- Provide legal backing to all forms of community institutions—crafted or drafted.
- Memorandum of Understanding (MoU) to be implemented after formalising all institutions.
- Equitable sharing of responsibilities.
- Benefits sharing of major forest produce as well.
- Changes in the functioning style of Forest Department from authoritarian to nurturing.
- Allow space for local decision-making.

- Tenurial and legal security at local level.
- Ban on transfer of land for non-forestry purposes.
- Seeking compatibility with Panchayati Raj Institutions (PRI), Tribal Forest Act, 2006 and the various forest related legislations.

Tribal communities have been managing their forests for centuries and there is no reason that they would not be able to do it now. At the same time they provide us with adequate lessons for incorporating the principles in programmes like the JFM so that they could be more successful.

References

Acharya, H.P. (1989). "Jirel Property Arrangements and the Management of Forest and Pasture Resources in Highland Nepal", *Development Anthropology Network* 7(2): 16-25. Binghamton, USA: Institute for Development Anthropology.

Acheson, J. (1987). "The Lobster Fiefs Revisited: Economic and Ecological Effects of Territoriality in Maine Lobster Fishing", in Bonnie J. Mc Kay and James M. Acheson (eds.), *The Questions of the Commons: The Culture and Ecology of Communal Resources*. AZ: The University of Arizona Press.

———. (1989). "Where have all the Exploitings gone? Co-management of the Maine Lobster Industry", in F. Berkes (ed.), *Common Property Resources*. London: Belhaven.

Adams, W., D. Brockington, J. Dyson and B. Vira (2002). "Analytical Framework for Dialogue on CPR Management", *CPR Policy Paper* 1. Department of Geography, University of Cambridge.

Agrawal, A. (1995). "Population Pressure = Forest Degradation: An Over-Simplistic Equation", *Unasylva* 46(2): 50-58.

———. (1998). "Group Size and Successful Collective Action: A Case Study of Forest Management Institutions in Indian Himalayas", in *Forest, True and People Working Paper*, Forest Resources and Institutions. Rome: FAO.

———. (2001). *The Regulatory Community: Decentralisation and the Environment in the Van Panchayats (Forest Councils) of Kumaon*. (www.yale.edu website)

Agrawal, Arjun (1994). *I Don't Need it but you can't have it: Analyzing Institutional Conflicts between Farmers and Pastoralists*. Pastoral Development Network 36a. London: ODI. pp.36-55.

Agrawal, R. (2001). "'Lath'. Panchayats: Fading Away", *Economic and Political Weekly* 36(1).

Arnold, J.E.M. and J.G. Campbell (1986). "Collective Management of Hill Forests in Nepal: The Community Forestry Development Projects", National Research Council, Proceedings of the conference on *Common Property Resource Management*. Washington, D.C.: National Academy Press. pp. 425-454.

Arnold, J.E.M. (1991). "Community Forestry—Ten Years in Review", *FAO Community Forestry Notes* No. 7.

Ascher, W. (1995). *Communities and Sustainable Forestry in Developing Countries*. San Francisco, California: ICS Press.

Baffa, J.M., G. Yameoga, P. Nikiema and D.M. Knudsen (1996). "Shea nut (Vitellaria Paradoxa) Production and Collection in Agro Forestry Parklands in Burkina Faso", in R.R.B. Leakey, A.B. Tamu, M. Melnyk and P. Vantomme (eds.), *Domestication and Commercialisation of Non-timber Forest Products in Agro Forestry Systems: Non Wood Forest Products* 9: 110-121. Rome: FAO.

Basu, N.G. (1984). "Community Forestry and the Local Community", in *Strategies and Designs for Afforestation, Reforestation and Tree Planting*, proceedings of an international symposium on the occasion of *100 Years of Forestry Education and Research in the Netherlands*, Wageningen, 19-23, September 1983. Wageningen, Netherlands.

Berkes, F. and C. Folke (1998). *Linking Social and Ecological Systems: Management Practices and Social Mechanisms for Building Resilience*. UK: Cambridge University Press, Cambridge.

Berkes, F., I. Davidson-Hunt, K. Davidson-Hunt (1998). "Diversity of Common Property use and Diversity of Social Interests in the Western Himalaya", *Mountain Research and Development* 18(1): 19-33.

Berkes, F., P. George and R.J. Prestpen (1991). "Co-management: The Evolution in Theory and Practice of the Administration of Living Resources", *Alternatives* 18(2).

Bhattacharya, P. (1995). "Emergence of Forest Protection by Communities in Kudada, South Bihar, India", *Community Forestry Case Studies Series* CM00295. Bangkok: Asia Forest Network and RECOFTC.

———. (1997b). "Learning Community Forestry from Nepal Experiences", *Journal of Tropical Forestry* 13(IV): 200-210, October-December.

———. (1997c). *Indigenous Approach of Managing Regenerating Forest Patch: Experiences from Bihar, India* (to be published).

Bhattacharya, P. and B. Joshi (2000). "Participatory Vegetation Assessment and Monitoring: A Tool for Assessing the Ecological Impact of Joint Forest Management", Paper presented at the *National Seminar on Sustainable Forest Management through Peoples participation* held at TFRI Jabalpur, 7-8 February.

Bhattacharya, P. and R. Prasad (1998). *Integrated Options for Forest Management in India*. Bhopal: Indian Institute of Forest Management.

———. (1998). "Integrating Forest Management Options through Local Resource Use Perceptions: Experiences from India", International Seminar on *Cultivating Forests: Alternative Forest Management Practices and Techniques for Community Forestry*. 23-25 September. Bangkok (Thailand).

Brightman, R. (1987). "Conservation and Resource Depletion: The Case of Boreal Forests Algongguiar", in Bonnie J. Mckay and James M. Acheson (ed.), *The Question of the Commons: The Culture and Ecology of Communal Resources*. Tucson, AZ: The University of Arizona Press. pp. 121-141.

Bruce, J.W. (1998). "Legal Bases for the Management of Forest Resources as Common Property", *Draft Community Forestry Note*. Rome: FAO.

———. (1990). "Rapid Appraisal of Land and Tree Tenure", *FAO Community Forestry Notes* No. 5.

Buchanan, J.M. and G. Tullock (1962). *The Calculus of Consent, Logical Foundations of Constitutional Democracy*. Ann Arbor: The University of Michigan Press.

Buchy, M. and S. Subba (2003). "Why is Community Forestry a Social and Gender Blind Technology? The Case of Nepal", *Gender, Technology and Development* 7(3). New Delhi: Sage Publication.

Chamberlin, J.R. (1974). "Provision of Collective Goods as a Function of Group Size", *American Political Science Review* 68(2): 707-16.

Chhetri, R. and T. Pandey (1992). *User Groups Forestry in the Far-western Region of Nepal*. Kathmandu, Nepal: ICIMOD.

Dasgupta, S. (2001). "A Note on Improvement in Regeneration Status in Forests of MP under JFM", *Indian Forestry*, July.

Davis-Case, D. (1989). "Community Forestry: Participatory Assessment, Monitoring and Evaluation", *Community Forestry Notes* No. 2. Rome: FAO.

Deb, D. and K.C. Malhotra (1997). "Interface Between Biodiversity and Tribal Cultural Heritage", *Journal of Human Ecology* 8:157-163.

Dinda, S. (2004). "Environmental Kuznets Curve Hypothesis: A Survey", *Ecological Economics* 49: 431-455.

FAO (2000). *Community-based Tree and Forest Product Enterprises: Market Analysis and Development*. Rome: FAO.

Feeney, D., F. Berkes, B. McKay and J. Acheson (1990). "The Tragedy of Commons: Twenty-two Years Later", *Human Ecology* 18(1): 1-19.

Feeny, D.H. (1988a). "Agricultural Expansion and Forest Depletion in Thailand, 1900-1975", in Richards J.F. and Tucker R. (ed.), *World Forest in the Twentieth Century*. Burham: Duke University Press, pp. 112-143, 281-287.

Fisher, R.J. (1989). "Indigenous Systems of Common Property Forest Management in Nepal", *Working Paper* No. 18. East West Centre, Honolulu: East-West Environment and Policy Institute.

Forest Survey of India (FSI) (2003). *State of Forests Report 2001*. Dehradoon: Forest Survey of India.

Gadgil, M. and R. Guha (1992). *This Fissured Land: An Ecological History of India*. New Delhi: Oxford University Press; Berkeley, CA: University of California Press.

Gautam, A.P. (2002). "Forestland Use Dynamics and Community Based Institutions in a Mountain Watershed I Nepal: Implications for Forest Governance and Management", *PhD Dissertation*, AIT Bangkok, Thailand.

Ghate, R. (2002). "Saga of Saigata", *Wasteland News* XVII(2): 1024-27.

Grahn, Richard (2008). "The Paradox of Pastoral Vulnerability", in Duncan Green, *From Poverty to Power: How Active Citizens and Effective States Can Change the World.* Oxfam International.

Gujit, I. (1999). *Participatory Monitoring and Evaluation for Natural Resource Management and Research.* University of Greenwich: Natural Resources Institute.

Hakkila, P. (1982). "Wood as a Worldwide Fuelwood Souce", in W. Ramsay Smith (ed.), *Energy from Forest Biomass*. London: Academic Press.

Hobley, M. and K. Shah (1996). *What Makes a Local Organisation Robust? Evidence from Nepal and India.* London: Overseas Development Institute.

Jewitt, S. (2002). *Environment, Knowledge and Gender: Local Development in India's Jharkhand.* Ashgate, USA.

Kant, S. (2000). "A Dynamic Approach to Forest Regimes in Developing Economies", *Ecological Economics* 32:287-300.

Khare, A., M. Sarin, N.C. Saxena, S. Palit, S. Bathla, F. Vania and M. Satyanarayana (2000). *Joint Forest Management: Policy, Practice and Prospects*. India: WWF.

Kumar, S. (2001). "Indigenous Community's Knowledge of Local Ecological Services", *Economic and Political Weekly* 36(30): 2859-869.

———. (2004). "Does Participation in Common Pool Resource Management Help the Poor—A Social Cost Benefit Analysis of JFM in Jharkhand, India", in Stuart Corbridge, Sarah Jewitt and Sanjay Kumar (eds.), *Jharkhand: Environment, Development and Ethnicity*. New Delhi: Oxford University Press.

Mayers, J. and S. Bass (2004). *Policy that Works for Forest and People—Real Prospects for Governance and Livelihoods*. Earthscan, London: International Institute for Environment and Development.

McKay, B. (1988). "Muddling through Clam Beds: Co-operative Management of New Jerseys Hard Clam Spawner Sanctuaries", *Journal of Shellfish Research* 7: 327-340.

McKean, M. (1992a). "Success on the Commons: A Comparative Examination of Institutions for Common Property Resource Management", *Journal of Theoretical Politics* 4(3): 247-282.

———. (1992b). "Management of Common Land (Iraichi) in Japan", in Daniel Bromley (ed.), *Making the Commons Work: Theory Practice and Policy*. San Francisco, CA: Institute for contemporary Press. pp. 63-98.

MoEF and Winrock India (2005). Workshop Proceedings on "National Consultative Workshop on Joint Forest Management (JFM)"—held in New Delhi between 14-15 July 2005. New Delhi: Winrock India.

Mukherjee, N. (1995). "Forest Management and Survival Needs-community Experiences in West Bengal", *Economic and Political Weekly* 30(49): 3130-132, December 9.

Naik, G. (1997). "Joint Forest Management: Factors Influencing Household Participation", *Economic and Political Weekly* 32(48): 3084-089.

Oliver, P.E. and G. Marvell (1988). "The Paradox of Group size in Collective Action: A Theory of the Critical Mass II", *American Sociological Review* 53(1): 1-8.

Ostrom, E. (1990). *Governing the Commons: The Evolution of Institutions for Collective Action*. New York: Cambridge University press.

Overseas Development Institute (1992). "Forest Policies and Forest Politics", in *ODI Agricultural Occasional Paper* No. 13, (ed.) by Gill Shepherd. London: ODI.

Pattnaik, B.K. and S. Dutta (1997). "JFM in South West Bengal: A Study in Participatory Development", *Economic and Political Weekly* 32(50): 3225-232.

Pearson, J.D. and D.D. Laitin (1996). "Explaining Interethnic Cooperation", *American Political Science Review* 90(4): 715-35.

Pecorino, P. (1999). "The Effect of Group Size on Public Good Provision in a Repeated Game Setting", *Journal of Public Economics* 72:121-134.

Pendzich, C., G. Thomas and T. Wohigenant (1994). "The Role of Alternative Conflict Management in Community Forestry", *Forest, Trees and People Working Paper* 1, Rome: FAO.

Pinkerton, E. (1987). "Intercepting the State Dynamic processes in the Assertion of local Comanagement rights", in Bonnie J. Mckay and James M. Acheson (eds.), *The Question of the Commons: The Culture and Ecology of Communal Resources*. Tucson, AZ: The University of Arizona Press. pp. 344-369.

Place, F. and K. Otsuka (2001). "Population Tenure, and Natural Resource Management: The Case of Customary Land Area in Malawi", *Journal of Environment Economic and Management* 41: 13-32.

Poffenberger, M. (ed.) (1990). *Keepers of the Forest: Land Management Alternatives in South East Asia*. USA: Kumarian Press.

Poffenberger, M., P. Bhattacharya, A. Khare, A. Rai, S.B. Roy, N. Singh and K. Singh (1996). *Grassroots Forest Research Network Report* Number 7. March. Berkeley: University of California, Centre for South Asia Studies.

Potetee, R.A. and E. Ostrom (2004). "Heterogeneity, Group Size and Collective Action: The Role of Institutions in Forest Management", *Development and Change* 35(3): 435-461.

Puri, E. (2004). "Understanding Participation Theoretical Foundations and Practical Applications", *Economics and Political Weekly* 39(25), June 12.

Ramam, S.S. (1981). *Dynamic Properties of Forest Ecosystems,* in D.E. Reichle (ed.), pp. 595-597. Cambridge, London; New York; Melbourne: Cambridge University Press.

Rasmussen, L.N. and R. MeinzenDick (1995). "Local Organisation for Natural Resource Management: Lessons from Theoretical and Empirical Literature", *EPTD Discussion Paper* No. 11. Washington DC: IFPRI.

Ravindranath, N.H., K.S. Murali and K.C. Malhotra (2000). *Joint Forest Management and Community Forestry in India*. New Delhi: Oxford and IBH Publishing Co. Pvt. Ltd.

Sarin, M., N.M. Singh, N. Sundar and R.K. Bhogal (2001). *Devolution as a Threat to Democratic Decision Making in Forestry? Findings from Three States in India*. Indonesia: CIFOR.

Singh, K. and V. Ballabh (1991). "People's Participation in Forest Management: Experience of Van Panchayats in U.P. Hills", *Wastelands News* VII(1).

Thomas, J.T. and C. Coulibaly (1995). "Common Property Forest Management Systems in Mali: Resistance and Vitality Under Pressure", *Unasylva* 180 (46).

Thompson, E.P. (1993). *Customs in Commons: Studies in Traditional Popular Culture*. New Press.

Thomson, J.R., D.H. Feeney and R.J. Oakerson (1986). "Institutional Dynamics: The Evolution and Dissolution of Common Property Resource Management", National Research Council, proceedings of the conference on *Common Property Resource Management*, Washington D.C.: National Academy Press, pp. 392-429.

Wade, R. (1988). *Village Republics: Economic Condition for Collective Action in South India*. Cambridge: Cambridge University Press.

World Bank (2006). *India: Unlocking Opportunities for Forest-dependent People*. New Delhi: Oxford University Press.

5

Jhum Cultivation: A Threat to Forestry?

MOIRANGTHEM PRAKASH

It is very unfortunate if Jhum only is blamed for denuding of forest in the North Eastern states of India. On the contrary, forests are exploited for timber; and hills are flattened for soil and stones. The basic philosophy of shifting cultivation has been to create forests and not to destroy forests, for without forests the next Jhum cannot be taken place.

With increasing convolution of global warming and mounting severity of climate changes, criticism against Jhum[1] is increasing day by day. Jhum cultivation is always blamed for its alleged role of deforestation and its consequent climate changes.

Shifting cultivation (jhuming practice,) which is directly linked with deforestation, is one of the main causes for extreme weather changes in the state (Manipur) despite steps being taken to wean the practices of the same by the state forest department. This is amidst the environmentalists' projection of increased landslides, floods in the state and other northeastern states by the impact of global warming.[2]

Jhum is blamed as ruinous and wasteful. It dries up the springs of the hills; causes soil erosion; destroys valuable forests; and adversely affects rainfall. Being frightened with these assumed adverse consequences, Central and concerned state governments initiated programmes aimed at arresting and eradicating shifting cultivation in the northeast. Contrary to this widely held notion about Jhum, there is another view, which is more liberal and humane. The latter view claims that Jhum is environment-friendly and sustainable. It considers the practice as an organic response of the people engaged in it to a certain specific ecological condition, rather to a particular ecotechno system (Bhowmik, 1980).

1. 'Jhum' is an Assamese term use commonly in northeast India. Every tribe in the region has its own name for the practice. For instance Tangkhuls call it 'Ahanglui' and 'Pam lou' by Meiteis.
2. Government of Manipur, Environment and Ecology Wing, Department of Forest, 'Jhum cultivation major culprit behind extreme environment changes in the state', February 9, 2009, on *www.kanglaonline.com*

While the hill people have to depend on Jhum cultivation for sheer sustenance, for survival, and therefore cause some amount of deforestation, we raise a hue and cry: they are destroying the forest ! But when contractors destroy the forest for collecting timber for furnishing quarters of gentlemen, for their multi-storied buildings, for providing slippers for railways used by them, we miss the fact that deforestation is caused by these contractors as well (ibid). In fact more than 60 per cent of India's wood products used such as plywood, sawn logs, wooden planks, matchsticks etc., came from this region prior to Supreme Court ban order on December 1996 (Dollo, 2004). If Jhum is responsible for deforestation and other climate changes, then to what extent Jhum cultivation is responsible for these? While Jhum cultivation has been practised by hill people through the centuries, did it create any ecological imbalance? Or, are we facing the crisis of ecological imbalance caused by deforestation by contactors sanctioned by governments? Why are the various measures adopted by Central and respective state governments to refrain Jhum have limited success?

The chapter makes an attempt to address these questions. Explaining how common property—land—is managed by hill tribes through this cultivation and to verify the fact that this form of cultivation is very much sustainable and environment-friendly are the basic thrust of this paper. The analysis of merits and demerits of Jhum will be an imperfect one without knowing properly the mode of Jhum cultivation; the management of common resources under this cultivation; and the close relationship of this practice with their (tribals') sociocultural life. This is dealt in the first section of the chapter. The second section addresses the two diverse views on Jhum; and the various measures adopted by the Central and the state governments and their limited success is discussed in the section three. The study concludes with looking forward to take up measures in consultation with local people and also keeping in mind unavoidable circumstances like increasing pressure on land due to population growth that leads to shortening of the Jhum cycle.

I

What is Jhum?

Shifting cultivation, regarded as a first step in transition from food gathering and hunting to food production, is nearly a 9,000 years old practice (Sharma, 1980: 4). In this type of land use system, farmers usually select a plot on hill slopes and clear the field for planting in part by slashing the vegetation and burning the debris. Thus, Jhum is sometimes known as slash and burn agriculture. On the cleared land the farmers grow crops only for few years—twice or thrice and then the plot is shifted to another place due to loss of fertility of the soil. The old one is left for years to recuperate. Cultivation is done with a very 'appropriate technology' such as machetes, sticks, hoes, or ling knives which are least capital-intensive. No animal is employed, there is no irrigation system. Clearance of the forest by burning process provides the necessary nutrients for the soil. In this farming practice there is rotation of fields and the crops are sown in calculated intervals, often between other plans, so the harvest can be staggered to provide food all the year round (Das, 2006).

This type of farming practice is popularly known as 'Jhum' in northeast India.[3] It is also known by various names in different parts of the country, for instance, Podu, Dabi, Koman and Bringa in Orrisa, Kumari in Western Ghats, Watra in south-eastern Rajasthan, and Pen du, Bewar or Dahia and Deepa in different parts of Madhya Pradesh (Das, 2006). This form of agriculture is practised mainly in regions which have relatively high temperature and abundant rainfall and in areas where the topography is not suitable for settled farming. The shifting cultivation is prevalent in different parts of the world, especially in Sumatra, northern Burma, New Guinea and in many parts of the African continent (Bhowmik, 1980). In India it is practised predominantly in the hill tracts of northeastern India, where the climate is monosonic with a high rainfall of over 2,200 mm followed by a dry winter and a brief warm summer, supporting a mixed subtropical humid forest. According to the estimates made by Task Force on shifting cultivation in India there were approximately 6,22.2 thousand families practising shifting cultivation in the country. Out of these, as many as 4,43.3 thousand families, constituting 71.25 per cent of the country, were located in the northeastern region alone (GoI, 1983).

How it Operates

There are over 100 different tribes in the northeast, differing linguistically and culturally. All these tribes, more than 6,20,000 families in the 7 states of northeast India practise Jhum and depend on for their subsistence (Ramakrishnan, 1992). According to Ramakrishnan, realising that there are over a 100 tribal communities which are highly insulated because of language, topography and sociocultural factors, one could reasonably expect over a 100 different variations in this land use practice. If ecological conditions are superimposed on this, then the heterogeneity gets very complex indeed. Therefore, it is important to realise that we are talking of a variety of Jhum

3. It is also known by different names in different regions. For example, it is called Berbaco and Coamile in Mexico; Conuco in Venezuela; Derrubadas, Quemadas and Oca in Brazil; Milpa in Central America; Chitenens and Mosale in Central Africa; Tavy in Malagasy; Ladang in Malysia; Djuma in Sumatra; Huama in java; Kiangin in Phillipines; Tamari in Thailand; Ray in Laos; Tuangin in Burma; and Chena in Sri Lanka.

systems rather than a single, homogeneous system of shifting agriculture, as often assumed in the scientific literature.

Although the pattern and details of shifting cultivation differ in different places and tribes, there is a remarkable uniformity of the methods of cultivation. The process of cultivation can be sketched as follows:

During the winter months (December-January), the undergrowth is slashed; and small trees and bamboos are felled. Short tree stumps and large tree boles are left intact. The underground organs of different species are not disturbed. This labourious effort is done by men from 2 to 3 families. Such a joint effort is one of the essential ingredients of a well-knit social organisation. This effort, along with the process of allotment of sites of Jhum by village headman who is in overall control of the village community, helps to promote kinship among the members of the village, and among many communities such as the Garos and the Mikirs (Maikhuri and Ramakrsihnan, 1990a).

Towards the end of March or the beginning of April, before the onset of monsoon, the debris is burnt *in situ*. Before burning, a fire line is cleared around the field. Burning is often repeated to destroy any unburnt material that has first been collected in heaps. A bamboo hut is built for temporary living. The family's presence protects the field from wild animals. Sowing is done after the first few monsoon showers. The seed mixtures used for different Jhum cycles may vary considerably. Whereas cereals are emphasised under long-Jhum cycles, perennials and tuber crops are emphasised under short-Jhum cycles, as by Garos in Meghalaya (Toky and Ramakrishnan, 1980). Some 8 to 35 crop species are grown together. Seeds of pulses, cucurbits, vegetables and cereals are mixed with dry soil from the site to ensure their uniform distribution, and broadcast soon after the burn. Maize seeds are dibbled at regular intervals amongst other crops. Similarly, rice is sown into the crop mixture by dibbling with a long stick, after the first rainfall in mid-April. Semi-perennial and perennial crops such as ginger, colocasia, tapioca, banana and castor are sown intermittently throughout the growing season. The leaves of *ricinus communis* are used for rearing young silkworm caterpillars. Weeding is a continuous process. Crops are harvested as they mature making room for the other crops left behind (Maikhuri and Ramakrsihnan, 1990b). Harvesting starts from the third month and continues upto December next. Thus a complete cycle of activities is completed for the first-year field.

Extent of Jhum

As discussed earlier, Jhum is the dominant agricultural production system in the mountainous tract of northeast India which has been practised since pre-historic days. Yet, surprisingly not much is known about the extent of Jhum in the region. No systematic survey has ever been carried out by any agency in the country to access the extent and status of Jhum cultivation. This is inspite of its great appeal and intense debate in the academic and administrative circles about the advantages and disadvantages of Jhum cultivation (Maithani, 2005).

The first attempt at estimating the extent of Jhum cultivation in the country was made by scheduled tribes and scheduled castes commissioner in his report of 1960-61. According to that estimate, about 25.89 lakh tribal populations were practising shifting cultivation on 5.41 lakh ha in the country. Of these, 18.56 lakh people were engaged in the practice on 3.32 lakh ha in the northeast India alone (Sharma, 1980). Subsequently the National Commission on Agriculture (NCA) based on the study conducted by Indian Council of Agricultural Research (ICAR) in 1965 made an attempt to estimate total number of tribal population, number of families and total area under shifting cultivation. Consequently, the Task Force on shifting cultivation constituted by the Government of India, Ministry of Agriculture also made an assessment of the extent of shifting cultivation in 1983. The Task Force reviewed the studies and estimates made by different individuals and institutions regarding shifting cultivation and related problems. According to the Task Force, the total area under shifting cultivation in the region was 13,515 sq km constituting 5.3 per cent of the geographical area of the region. And annually 3,871 sq km area was subjected to shifting cultivation by about 4,43,300 families. The area cultivated annually constituted only 1.5 per cent of the geographical area of the region (GoI, 1983). The following table highlights status of Jhum cultivation in the northeast.

Table 5.1

Shifting Cultivation in the Northeastern States

State	*Annual Area Under Shifting Cultivation (sq km)*	*Fallow Period (in years)*	*Minimum Area Under Shifting Cultivation One Time or Other (sq km)*	*No. of Families Practising Shifting Cultivation*
Arunachal Pradesh	700	3-10	2,100	54,000
Assam	696	2-10	1,392	58,000
Manipur	900	4-7	3,600	70,000
Meghalaya	530	5-7	2,650	52,290
Mizoram	630	3-4	1,890	50,000
Nagaland	190	5-8	1,913	1,16,046
Tripura	223	5-9	1,115	43,000
Total	3,869 (1.5 per cent)		14,660 (5.7 per cent)	4,43,336

Source: Basic Statistics of NER, 2002, Government of India, North Eastern Secretariat, Shillong. p.42.

The annual area under shifting cultivation in the region is 3,869 sq km and 4,43,336 families from the region practise Jhum. Manipur, though tiny in size, has the largest size in area—900 sq km—under shifting cultivation annually among the northeastern states. As much as 70,000 families from the state engage in shifting cultivation. There are 1,16,046 families in Nagaland who rely on this cropping pattern, though the annual size under shifting cultivation is only 190 sq km.

Ownership and Management of Land

One of the essential characteristic features of shifting cultivation regime is community control over the land. There are varieties of traditions of land use but the common feature that permeates the whole region is that the land is tenure free. Under Jhum, land is a common property, which is collectively controlled and owned by the village community. The community as a whole selects the plot and decides for cultivation. The chieftain or village head designates land to the individual families for their use on the basis of the family size. There was no landless class, no land speculation and exploitation on the basis of private ownership, and the fertility of the land was maintained. Traditionally, each village had its well recognised territory within which all the members/families enjoyed the right to cultivation, hunting, fishing, gathering etc. In some tribes the territory is divided between clans and further down to lineages of the clans which gives an impression of *de facto* private ownership. In such cases also, only the right to use is enjoyed by members undisturbed and continuously but they cannot sell the land at their will though it is inheritable. There is no written law on the land system except the shifting cultivation regulations based on customary and traditional practices which recognise only village as an entity and community as owner of the village land. Thus, the land system which varies in details from tribe to tribe has one common feature that the land is tenure free with individual rights of occupation and communal rights of alienation (Maithani, 2005).

Here we can take up the case of Arunachal Pradesh to have a deeper understanding of how common property—land—is managed under this farming practice. Dollo, in his study 'Shifting cultivation and conservation laws of Arunachal Pradesh', made an attempt to examine in detail the land tenure system in the state, which is a unique one (Dollo, 2004). The land ownership of all the tribes shows a well demarcated area under its jurisdiction and is administrated by a traditional village council (TVC) headed by a chief or headman or by village secretary and president in some cases. The local customary laws that govern the land tenure systems have evolved over long experimentation, which are tribe specific where each tribe or even village has its own unique customary laws and traditions. All the indigeneous community of the state have traditionally independent unit of administration of their own. The village councils, chiefs or headmen are generally custodian of the land, where all members of the community have a right to use it freely but with consent of the custodian. As each family member of the village or community is given a plot(s) of land, and the family has the right to cultivate until the village community decides to move away from that area to somewhere else where more virgin forestland for cultivation and easy approach to metallic road is available. In early history, authority of distribution of land for farming was vested on the headmen or village but since the last decade, trend has slightly changed by first selection of the land by a farmer. No part of the community or village land can be sold to any one by the chief or the village council without the consent of the village people and is totally restricted on the use of land by outsiders. Although, the land permanently developed and owned by individuals can be transfered as per his convenience, he can only sell it to a person belonging to that village or community itself. In case of disputes in land tenureship or ownership, the traditional village council such as *mangijombana, jung-me, melley, nimiyang* among Monpa, Sherdukpen, Aka and Bugun, respectively or village headmen (*gaon bura*) resolves the disputes and their decisions are binding and final (Dollo, 2004).

It has been considered as a healthy feature of the tribal land system which guarantees access to land for cultivation to all its members and protects them against land alienation. It is significant for maintaining equity of economic and social status and security of livelihood to the members of the community. In appreciation of this healthy custom, tribal communities of the northeast India have been granted safeguards under the Sixth Schedule of the Indian Constitution and several other provisions to protect their rights on land and to preserve their cultural and customary practices of self-management. In the absence of the codification of customary rights and laws, the prevailing land tenure arrangements are very complex and at times unclear. However, households are not prevented from taking up permanent cultivation on the land allotted for shifting cultivation. This of late has paradoxically led to quite privatisation in flagrant violation of customs resulting in concentration of land in few hands at the cost of the survival needs of the large number of Jhumias (Maithani, 2005).

Division of Labour and a Way of Life

Labour is carried out by the family and is designated according to a culturally specific division of labour. Usually, women predominate in seed selection, dibbling, planting and weeding, while other activities such as clearing jungles and burning of the cut undergrowths are carried out by their counterparts. On the day of sowing which is a ceremonial day for the whole village, it is interesting to observe the male members of each family on reaching the Jhum field in the early morning engage in preparing the digging sticks. The male members sow seeds by broadcasting. Both men and women participate in harvesting. Such a joint effort is one of the essential ingredients of a well-knit social organisation. The process of allotment of sites of Jhum by the village headman, who is in overall control of the village community, helps to promote kinship among the members of the village, and among many communities.

The whole gamut of primitive society is interwoven with the means of food production, which in this case, is Jhum cultivation. Their way of life, training of youth, the whole social and political system, the ceremonies and festivals, in other words, their philosophy of life is a product of this system of the economy (Sharma, 1980). This practice is not simply a form of cultivation; rather it is a way of life for traditional tribal societies which they have been practising since pre-historic times. Jhum plays an important cultural role in local customs, traditions and practices, besides offering economic security to farmers.

II

As we discussed in the beginning, there are two diverse views on Jhum. One view blames Jhum as ruinous and wasteful as it directly leads to deforestation; causes soil erosion; dries up the springs of the hills; and adversely affects rainfall. Contrary to this widely held notion about Jhum, there is another view, which claims that Jhum is environment-friendly and sustainable. The debate on advantages and disadvantages of Jhum have been going on for more than last 50 years among academics, environmentalists and in the administrative circle.

What is Wrong with Jhum?

In the past when a balance was maintained between population and soil fertility as a result of a longer fallow cycle of 10 to 20 years, Jhum enabled multiple cropping of several crops which provided a balanced diet and also offered some form of crop insurance to the Jhumias in the event of failure of some crops. It helps to conserve forests as the pattern observes a long fallow period followed by a short cropping phase. In the recent past Jhum cultivation has been given a bad name, allegations have been made and even measure have been taken up by the Central and concerned state governments, though met with limited success, to restrict this practice.

Classifying Jhum

The practice of Jhum can be classified into different types. This classification aids in understanding how Jhum has changed over the years and what is going wrong with newer practices of Jhum. According to B.K. Tiwari, Jhum can be broadly classified as traditional, distorted, innovated and modified (Sule, 2006).

1. Traditional shifting agriculture is mostly found in villages which have not experienced much pressure of population increase. It is sustainable but may not fulfill all the needs of a modern society. It helps to conserve forests as the pattern observes a long fallow period followed by a short cropping phase.
2. In land scarce villages, for reasons such as population increase, fallow periods are reduced and shifting agriculture has got distorted. Elsewhere the Jhum has spread to lands with steep slopes due to non-availability of lands with lesser degrees of slopes. In Mizoram, for instance, at places less than 1000 metre altitude, the distorted shifting agriculture has converted much of the subtropical evergreen and semi-evergreen forests into scrub and bamboo brakes.
3. In the third type, i.e., innovated shifting agriculture, farmers have switched to newer methods of cultivation which are more adapted to the current availability of resources (for example green biomass, technology and labour) and present day societal requirements (for example cash). In some cases they have also introduced new crops, (for example Kolar Beans in Nagaland, Pisum in Meghalaya) which help maintain fertility and also have good market.
4. The fourth type—modified shifting agriculture—got introduced during the past decade with implementation of two development projects *viz.*, Nagaland Empowerment of People through Economic Development (NEPED) in Nagaland, and North Eastern Region Community Resource Management Project (NERCORMP) in Meghalaya, Manipur, and hill districts of Assam.

Over the years, many tribal societies got pushed into employing distorted versions of shifting agriculture. In land scarce villages, for reasons such as population increase and consequent increase in demand for food, fallow periods are reduced. Greater chunks of forest lands have cleared to grow more food and returned to the fallow plots much sooner than 10-20 years. The length of the fallow period between two successive cropping phases has come down to even 2 to 3 years in some places. This has resulted in soil degradation; fall in yield; lower returns; and reduction in green cover.

Deforestation, Soil Erosion and Ecological Disturbance

It is blamed that Jhum directly leads to deforestation; causes soil erosion and disturbs ecological biodiversity. According to Stracey, this 'primitive' form of agriculture results in serious environmental problems: loss of forest cover, erosion of topsoil, desertification, and declines in forest productivity (Stracey, 1967). It has also decried Jhum as an inefficient form of agriculture, an impediment to progress of forestry, and an agent of destruction of biodiversity (Borah and Goswami, 1973). Such beliefs have been widespread since British times, and have even resulted in forcible suppression of the practice, oppression and relocation of tribals in Central India and other hill regions (Von 1982).

According to ENVIS[4] Centre, Manipur, increase in area and intensity of Jhuming and also decrease in the Jhum cycle due to high population growth are destroying the forests and the rich biodiversity, making it on the verge of becoming totally degraded due to loss of top soil, fertility and vegetative cover ultimately turning into a barren wasteland.

Frequent shifting from one land to another has affected the ecology of the region. The area under natural forest has declined; the fragmentation of habitat, local disappearance of native species and invasion by exotic weeds and other plants are some of the other ecological consequences of shifting agriculture. The area having Jhum cycle of 5 and 10 years is more vulnerable to weed invasion compared to Jhum cycle of 15 years. The area with 15-year Jhum cycle has more soil nutrients, larger number of species, and higher agronomic yield with ratio of energy output to input as 25.6 compared to Jhum cycle of 10 and 5 years (Misra and Ramakrishnan, 1981). Similarly, while studying Jhum ecology in Meghalaya, it was reported that water and nutrient losses in shifting cultivation areas were far greater than in the virgin areas, and areas left for 50 years after Jhuming (Singh and Ramakrishnan, 1982). Thus, reduction in the cycles of podu, or Jhuming, adversely affects the recovery of soil fertility, and the nutrient conservation by the ecosystem. Repeated short-cycle Jhuming has created forest-canopy gaps which are evident from the barren hills.

Dilution of Customary Practices

The last 50 years have seen the decline of traditional institutions and dilution of customary practices under the impact of outside exposure and expansion of modern state machinery in these areas. As a result of this, tendency to acquire and accumulate private property in land has increased

4. Environment Information System, ENVIS Centre, Manipur. Environment and Ecology Wing, Government of Manipur.

taking advantage of loopholes in the custom. The shifting cultivation control programmes have further fuelled this tendency by emphasising change over to settled agriculture, and horticulture plantations. Wide ranging incentives including upto 100 per cent subsidy to villagers for converting Jhum field into terraces and orchards and many other facilities were offered by the government for this purpose. Not that shifting cultivation has been given up, but the well-placed, articulated and educated section of the people have been able to manipulate the custom to acquire ownership of land and other productive assets at 100 per cent government cost. This has given rise to disparities in income and assets and shrinking of common property resource base (shifting cultivation lands) rendering the poor more vulnerable. Although the state governments have not conferred legal titles to the holders of these assets but *de facto* private ownership of such lands is recognised not only by the communities but also the government which was instrumental in the creation of these assets in the first place (Maithani, 2005).

It is cleared that the problem with this cropping pattern is the distortion in the original version of Jhum. It is this change in traditional practice—increase in Jhum area and intensity of Jhuming with shortening Jhum cycle—arising out of the changing conditions, that has given Jhum cultivation a bad name.

Jhum is Favourable

In contrast to the above arguments against Jhum, there are various arguments that is strongly in favour of Jhum. Experts and environmentalists are of the strong opinion that Jhum is more environment-friendly and sustainable. Findings of various studies and significant body of literature also substantiate the opinion and lead to the conclusion that Jhum cultivation with longer cycles has many advantages over other agricultural systems in the hills.

Jhum Conserves Forest

One of the chief objections raised against Jhum is that it leads to deforestation, as the hill people who engage in shifting cultivation indiscriminately destroy the trees and the forest. It is very unfortunate if Jhum only is blamed for denuding of forests in the northeastern states of India. On the contrary, forests are exploited for timber and hills, are flattened for soil and stones. The charge of felling trees indiscriminately does not have a very solid foundation. Tribal people have a very close link with the nature, with the forest, and they too are lovers of trees. In fact these people draw their sustenance not only from shifting cultivation but from different kinds of forest produce. Naturally, they, for their own interest, would not destroy trees indiscriminately; they would save those trees from which they get fruits or other things which they can use for their own purpose (Bhowmik, 1980).

Jhum Minimises Soil Erosion

Those who subscribe to the opinion that Jhum results in soil erosion will find it disquieting that shifting cultivation is a response of the tribal people of the hill areas to the problem of erosion of fertile top soil of steep slopes. As M.S. Shivaraman, Advisor to the Programme Administration of the Planning Commission, observed in 1957:

> It is a mistake to assume that shifting cultivation in itself is unscientific land use pattern. Actually, it is a practical approach to certain inherent difficulties in preparing a proper seedbed in steep hill slopes where any disturbances of the surface by hoeing or ploughing will result in washing away of the fertile top soil (Bhowmik, 1980).

The tribal people, therefore, take care not to plough or disturb the soil before sowing. The destruction of weeds and improvement of tilth necessary for a proper seedbed are achieved with the help of fire. Soil erosion is minimised in Jhum due to the retaining of rootstocks of bamboo and trees in burned plots, the rapid recovery of weeds and bamboo following abandonment, and the interspersion of forests and fields on hill slopes. The evidence for siltation of rivers and floods because of soil erosion due to Jhum is weak and possibly untenable.

Jhum is Sustainable and Preserves Biodiversity

It is commonly believed that Jhum was a sustainable system in the past when fallow periods were long, but with increasing population pressure Jhum cycles have declined and become unsustainable. The evidence for this is weak. In a review, Singh (Singh, 1996) has pointed out that: (i) there is often no strong relationship between population pressure and Jhum cycles, (ii) villagers choose to cultivate at cycles of 5-10 years even when longer fallow periods are possible, and (iii) that population density would impinge on Jhum cycle only after some critical threshold of high population pressure is crossed. The duration of the fallow period is influenced by the ease of clearing the vegetation and soil fertility levels following the slash-and-burn operations. The burning of slash returns nutrients to the soil through ash and kills microbes allowing relatively high yields. Yields decline as the soil is depleted through one year of cropping and cultivation is rarely carried out for more than a year. When fields are abandoned, there is rapid regeneration of bamboo and other plants. After 10 years, the vegetation and soil properties recover to levels that can support another round of Jhum cultivation (Ramakrishnan, 1992). In some areas with better regrowth of bamboo, even shorter fallow cycles may be feasible and sustainable (Singh, 1996). Where fallow cycles are less than 5 years, Jhum cultivation may be stabilised using modern technical inputs, while concurrently developing alternative village based occupations (Ninan, 1992).

Studies by ethnologists have tended to view shifting cultivation favourably. It is considered a diversified system, well adapted to local conditions in moist forest and hilly tracts (Conklin, 1969). Others have argued that traditional shifting cultivation may not be as destructive as modern forest exploitation for timber. Clearance of small patches of forest with long fallow periods may even enhance biodiversity in the landscape due to the creation of a variety of habitats (Gadgil and Guha, 1992).

Efficiency of Jhum

The economics and efficiency of shifting agriculture has been studied in Meghalaya and other states of northeast India by a team of scientists led by P.S. Ramakrishnan. These studies showed that, far from being primitive and inefficient, Jhum is an ingenious system of organic multiple cropping well suited to the heavy rainfall areas of the hill tracts. The economic and energetic efficiency of Jhum is higher than alternative forms of agriculture such as terrace and valley cultivation. This is mainly because terrace and valley cultivation needs expensive external input such as fertilisers (which often get leached or lost in the heavy rainfall at hill slopes) and pesticides, besides labour for terracing. Monetary output-input ratios (rupees/ha/year) range from about 1.8 for Jhum cycles of 5-10 years to 2.1 for a 20-year Jhum cycle. Even gross returns (rupees/ha) from Jhum are about 1.9 times higher than from terrace cultivation (Ninan 1992).

Superiority of Jhum

The superiority of Jhum cultivation over some forms of sedentary cultivation partly explains the persistence of this form of agriculture in northeast India. Other reasons include the economic security provided by Jhum and its cultural importance to indigeneous tribes. Poor access to markets, capital, and technical know-how of more commercially rewarding alternatives such as horticulture and cash crop cultivation also hinders the transition to other occupations. Clearly, one cannot do away with Jhum assuming it to be a primitive and inefficient system, as attempted in governmental Jhum control programmes and new land use policies (Singh, 1996). Instead, an unbiased understanding of the advantages of Jhum is required for proper design and implementation of developmental programmes.

III

Measures to Stop Jhum

Being frightened with the disadvantages brought out by present distorted version of shifting cultivation, Central and concerned state governments initiated programmes aimed at arresting and eradicating shifting cultivation in the northeast. Among the various programmes, mention may be made of 'Jhum regulation', 'Jhum control' and 'soil conservation' which was started during 5th and 9th Five-Year plans. The measures under these programmes include: a) resettlement of Jhum farmers by relocation of their villages and provision of alternate means of livelihood; b) introducing terrace cultivation; c) diversification into horticulture, floriculture and plantation crops; and d) a complete change in land tenure system in the hills whereby community—owned and operated holdings are replaced by individual holdings.

Various attempts have been made by governments to settle the tribals involved in shifting cultivation. For instance: (i) arable land is provided to the tribals for carrying out agriculture and also to settle in the area; (ii) an Agro-forestry project known as NEPED funded by Canadian International Development Agency (CIDA) through India–Canada Environment Facility (ICEF) was initiated in 1995 to make Nagaland self-sufficient in agroforestry. The objectives of the project were: (a) identification of trees by local tribes and demonstration of method leading to more sustainable resource management; (b) addressing people's need, and evolving better management systems; and (c) promoting marketing initiatives. Under this project, experienced government officials convince the villagers to set aside six hectares of land, known as test plot, in Jhum areas for over two and half years. Owner of this plot is required to plant 1,200 seedlings, along with usual crop, under supervision of village council and project team.

The Strategies

1. Providing employment opportunities and income generation on a regular basis through proper utilisation of the land resources, i.e., by equitable distribution of wasteland among the tribals.
2. By encouraging cooperative efforts for carrying out forest-based activities, i.e., basket making, rope making, cane furniture processing of minor forest produce, honey collection, etc., have to be made commercially viable by providing proper marketing facilities.
3. By forming village forest committees for protection and development of the degraded forests. These committees by providing suitable

incentives to the tribals, after the time of harvest can divert some of the tribals away from the shifting cultivation.

4. By ensuring implementation of total literacy campaign.

The ICAR also introduced a model of land use for the hilly areas in the northeastern region that consists of bench terracing for agriculture at the lower portion of the hill slopes. The mid-portion of the hill slope is to be converted into half-moon terraces for horticulture and the top portion for forestry.

Failure of the Schemes

Initially these programmes were based on the assumption that introduction of cash crops would help to wean people away from Jhum. Jhumias were given planting material and financial help to switch over to these crops. These schemes have, however, not yielded the desired results. The primary cause of the failure to attract people to settled farming was that the new settlements provided were not readily accepted by the people due to their close attachment to their traditional villages and way of life. Moreover, to switch to terrace cultivation or the use of bullocks for ploughing causes great technical difficulties for Jhum farmers. Limited availability of land for terrace farming at higher altitudes is also a major problem. Some terraces, built with retaining walls at a height of 3 to 7 feet, are very difficult to maintain during the rains, resulting in the collapse of whole terraces with standing crops. A prerequisite for the success of terrace farming is the development of a proper technology for water management and water conservation. Since terraces are above perennial streams and rivers, it is essential that adequate power is available for lift irrigation schemes. Thus, while promoting terrace cultivation on the model recommended by the ICAR, due consideration and attention needs to be given to all the above mentioned requirements. However, the prevailing pattern of community land ownership in the hill areas and the high capital investment involved in the ICAR model deter the Jhumias from adopting it (ICAR, 1990).

The ignorance of the authorities on the socioeconomic and agroecological conditions of shifting cultivation and the minimal involvement of forest department officials, who are more informed about the above factors, in implementation of schemes are also another important factor responsible for the failure of these schemes. There are some other causes also as being discussed below (Borthakur, 2002):

1. Non-involvement of the community and rural institutional infrastructure in decision-making and execution.
2. Undue stress on development of terraces and poor production in dry terraces devoid of top soil due to lack of irrigation.

3. Exclusion of entire villages or communities from the programme, in many cases.
4. Laying of stress exclusively on raising rice crop inducing farmers to revert back to Jhuming for raising the other crops.
5. Conversion of farmers of horticultural and plantation crop to wage earners without taking care of food security in remote areas.

If these programmes are implemented without massive public investment and instituting a more participatory model of development planning in consultation with the communities for whom they are designed, they are, likely to cause major disruptions in the agricultural systems, food security, and way of life of the hill people and possibly endanger the very identity of tribal societies and their traditional institutions. Thankfully they (farmers) defeated the effort of government to switch over to cash crops. If succeeded, we would have witnessed farmers' suicide among the tribals in the region in an unprecedented way.

IV

Beyond the Narrow Views

The failure of various government schemes and programmes; farmers' reluctance to switch over to other forms of agriculture—terraced and settled; and their continuance of traditional Jhum cultivation are clear testimony of supremacy of shifting cultivation over other alternatives in the hilly terrain region of northeast. At a conference[5] at Shillong, a Jhum farmer said, 'after all is said and done, whatever new scheme is brought to us, it is only Jhum cultivation which can sustain us and our family as we live in the steep mountain areas where terraced or settled cultivation is just not possible'. Jhum cultivators conserve forests and make it productive at the same time. Jhumias nurse and nurture forests back into their Jhum fields, while conventional farmers banish the forest from their fields. It was also a way of forest management at the landscape level. The basic philosophy of shifting cultivation has been to create forests and not to destroy forests, for without forests the next Jhum cannot take place.

It would be unfortunate if developmental programmes based on misguided opinions about Jhum suppress this unique form of agriculture. Only occupations providing monetary and social benefits perceived by Jhumias to outweigh the cultural and security benefits embodied by Jhum are likely to

5. Regional Shifting Cultivation Policy Dialogue Workshop for Eastern Himalayas, October 8-10, 2004, Shillong.

gain acceptance. A balanced approach to development that also recognises the merits of Jhum is needed. Then, this remarkable form of organic farming may persist into the 21st century.

A meaningful solution to the problem of Jhum has become critical, not only from the point of biodiversity conservation, but also for productive agriculture in the region. The solution to the problem of the shortening fallow cycle lies in strengthening this already weakened agroforestry system, through fallow management using appropriate tree species, rather than imposing an alien technology from outside the region. Nepalese alder (*alnus nepalensis*) and bamboos are examples amongst many others to be selected through intense community participation (Ramakrishnan, 1992).

The need for better regulation seems of utmost importance. Farmers should be allowed to clear the fallows when it is time for the cropping phase, and the species selection should be adjusted towards soil conservation as well. The timber trade should be regulated in such a way that the farmers can sell the timber when they clear it. Existing common property regimes should be strengthened, to avoid capture of the land and improved fallows by elites from within the communities. Overall, tenure security should be increased, as well as access to credit in those situations where common property regimes apply. Policies and programmes should be put in place that better acknowledge the role of customary institutions in the community management of the resources, in controlling the burning, and protecting forest resources against the timber mafia and elite capture.

Also, in recent years, the joint forest management (JFM) concept has come into vogue for combining the state and the people in the forest management. The current programme is not appropriate, however, because farmers are supposed to be the owners of these forests, whereas in JFM they would have to share ownership. Moreover, the forest departments focus on policing and controlling the forests. If they were more adept at training farmers and working with communities, they could play a major role in mobilising them to improve and enrich their fallows by growing forests there.

To conclude, emphasis should be on controlling distortions instead of controlling shifting agriculture itself. Alternative measures should be provided to meet the increasing demand for food while efforts are made to preserve the original form of cultivation. Proper consultation with the community, their full consensus and active participation are exceedingly necessary for whatever measures taken up in order to not to repeat the same failure story.

References

Bhowmik, P.K. (1980). "Shifting Cultivation: A Plea for New Strategies", in *Shifting Cultivation in North East India*. Shillong: North East India Council for Social Science Research. pp. 6-13.

Borah, D. and N.R. Goswami (1973). "A Comparative Study of Crop Production Under Shifting and Terrace Cultivation (a case study in the Garo hills, Meghalaya)", *Ad hoc Study* 35, Jorhat: Agro-economic Research Centre for North East India.

Borthakur, D.N. (2002). "Shifting Cultivation in Northeast India: An Approach Towards Control", in B.J. Deb (ed.), *Development Priorities in Northeast India*. New Delhi: Concept Publishing Company. pp. 123-134.

Conklin, H. (1969). "An Ethnoecological Approach to Shifting Agriculture", in A.P. Vayda (ed.), *Environment and Cultural Behaviour*. New York: Academic Press. pp. 221-233.

Das, Debojyoti (2006). "Demystifying the Myth of Shifting Cultivation: Agronomy in the North-east", in *Economic and Political Weekly* 47: 4912-916, November 25-December 1.

Dollo, Mihin (2004). "Shifting Cultivation and Conservation Laws of Arunachal Pradesh", in *North Lakhimpur Laws College Magazine* I: 43-45.

Government of India (1983). *Task Force on Shifting Cultivation*. Ministry of Agriculture.

———. (2002). *Basic Statistics of NER*. Shillong: North Eastern Secretariat, p.42.

Gadgil, M. and R. Guha (1992). *This Fissured Land: An Ecological History of India*. New Delhi: Oxford University Press.

ICAR (1990). *Technological Bulletin*—ICAR Research Complex for NEH Region during the decade (1975-1984). Vol-I: Crop Science, ICAR Research Complex for NEH Region, Shillong, Meghalaya.

Maikhuri, R.K. and P.S. Ramakrishnan (1990a). "Ecological Analysis of Cluster of Villages Emphasising Land Use of Different Tribes in Meghalaya in Northeast India", *Agriculture, Ecosystem and Environment* (32): 17-37.

———. (1990b). "Comparative Analysis of the Village Ecosystem Function of Different Tribes Living in the Same Area in Arunachal Pradesh in Northeastern India", *Agricultural Systems* 35(4): 377-99.

Maithani, B.P. (2005). *Shifting Cultivation in North-east India: Policy Issues and Options*. New Delhi: A Mital Publication.

Misra, B.K. and P.S. Ramakrishnan (1981). *Oecologia Applicata* 2: 369–389.

Ninan, K.N. (1992). "Economics of Shifting Cultivation in India", *Economic and Political Weekly* 27(3): 2-6. March 28.

Ramakrishnan, P.S. (1992). "Shifting Agriculture and Sustainable Development: An Interdisciplinary Study from Northeastern India", *MAB Series* 10. Paris: UNESCO.

Sharma, C. Tarun (1980). "The Prehistoric background of Shifting Cultivation", in *Shifting Cultivation in North East India*. Shillong: North East India Council for Social Science Research.

Singh, D. (1996). *The Last Frontier: People and Forests in Mizoram*. New Delhi: Tata Energy Research Institute.

Singh, J. and P.S. Ramakrishnan (1982). *Proc. Indian Acad. Sci.* B91. pp. 269–280.

Stracey, P.D. (1967). "A Note on Nagaland", *Journal of the Bombay Natural History Society* 64: 440-446.

Sule, Surekha (2006). "Jhum Cultivation under Sharper Scrutiny". *http://www.indiatogether.org/2006/dec/agr-Jhum.htm*

Toky, O.P. and P.S. Ramakrishnan (1980). "Cropping and Yields in Agricultural Systems of the North-eastern Hill Region of India", *Agro-Ecosystems* 7: 11-25.

Von Fürer-Haimendorf, C. (1982). *Tribes of India: The Struggle for Survival*. New Delhi: Oxford University Press.

6

Traditional Institutional Mechanisms for the Management of Common Property Resources

A Case From Karnataka

ASHOK S. ALUR, N.V. HIREMATH, R. MOHAMMED,
JANE CARTER, RUPA MUKERJI and B.N. DIVAKARA

Introduction

Tanks occupy a predominant place in the history of common property resources (CPRs) irrigating enormous area in South India. About 44,000 tanks serve 27,000 villages of Karnataka. The area irrigated by tanks during the pre-Independence era was 6,84,000 ha and has come down to 2,40,000 ha, and is declining year after year (Madangopal, 2002). Tanks are the structures built against a stream or valley for harvesting rainwater or the surface run off. The rocky tanks of South India are potentially best suited to recharge ground water and to extend safe drinking water to the families in the command area (Reddy, 1991). In spite of this, they have lost their significance over the last few decades due to: (a) emphasis of the governments on large scale irrigation projects and well irrigation (including borewells), over highly decentralised tanks (WCD, 2001) in the post-Independence period; (b) lesser allocation in the budgets for operation and maintenance of tanks (Reddy and Ananathramaiah, 2002); (c) introduction of superior water lifting technologies and rural electrification which has encouraged farmers to prefer private well irrigation to community based tank irrigation (Reddy, 1997); (d) take over of management of tanks of all sizes by the Minor Irrigation Department (MID) (Reddy and Ananathramaiah, 2002); (e) marginalisation of age-old social institutions constituted by the communities for tank management; and (f) alienation of the community for the outputs from tanks and subsequent transfer of such rights to respective departments. Despite all the above listed causes, there are documented and reported cases of tanks (Reddy, 1991; Sengupta, 1993), such as Mudiyanur of Mulbagal *taluk*, which are being effectively managed by communities as a source of irrigation. The community has a management institution which has effective practices to meet the irrigation needs of the farmers of the command in an equitable manner apart from protecting ground water levels.

Objectives of the Study

The study was planned and undertaken with the following objectives:

1. To understand the inherent strengths of the social institution for its survival under the prevailing social, economic and political conditions.
2. To examine the effectiveness of the practises adopted in water management for crop production and allocation of gains in terms of equity and social considerations.
3. To study the feasibility of continuing the working of the institutions under various socioeconomic and political conditions.
4. To understand the possibilities of adopting the principles, practises and other favourable institutional aspects of traditional institutional structure, within the new institutional structure framed by various government organisations involved in CPRs/natural resources management.

Location of Mudiyanur Tank

The Mudiyanur tank is located in the southeastern part of Karnataka in Kolar District, a semi-arid region along the eastern dry zone of the Deccan Plateau. The temperature in the area varies from 32°C in the summer to 20°C in the winter. The southwest monsoon brings precipitation from mid-June to mid-September and the average annual rainfall for 26 years stands at 787 mm with its peak during the end of July to mid-August. Rocky granite outcrops are common features in the area. The tank is situated between Mudiyanur and Bestrahalli village on the main road connecting Mulbagal town which is 15 km away from the tank whereas Kolar is around 45 km away. Mudiyanur tank is in the downstream of chain of tanks located in the watershed area of Uttanur. This tank is succeeded by the Uttanur tank, which empties the entire surplus from the chain of tanks into Palar River. Consequently, Mudiyanur tank succeeds several tanks of various sizes.

History

The history of Mudiyanur tank is not clearly known. As per information available it was built around 828 AD, and was the exclusive property of the *Jodidars* residing at Mudiyanur village. There were 56 families having rights to the *Jodidar* system and each family head took it in turn to be the head of the village, bearing the name *Patel*. As the tank was an exclusive property of

Jodidars, the entire command area was also exclusively owned by them. Only the *Neergantis* were given *inam* land. The villagers in the neighbourhood villages were hired in as tenants. After the abolition of the *Jodidar* system in 1957, the *inamathi* system in 1976 and subsequent land reforms in the 1970s, the *Jodidars* started selling their holdings within the command area to the farmers within the Mudiyanur village and other villages in the neighbourhood. Even today, about 40 per cent of the land in the command area is cultivated by tenants.

Table 6.1

Traditional Terms and their Meaning

Term	*Meaning*
Jodidars	*Jodidars* were appointed by the British rulers during colonial period to collect revenue and administer law and order in the villages.
Jodi	Village and the land within its administrative boundaries for which land tax had to be collected by the designated person (*Jodidar*).
Neerganti	Person identified by the community for allocation and distribution of water in the command area of a tank, and protection of crops in the command area from thieves, cattle, etc.
Inam land	*Inam* usually refers to the gift. In this case a piece of land given in lieu of the salary to those who were in the services of the British or the native king and located at the village level. This piece of land was inheritable.
Inamathi	This refers to the ownership of *inam* land.
Patel	Patel means the title of the village head. Traditionally, this was the position on rotation enjoyed by the *Jodidars*. Officially, the post of *Patel* is now abolished.
Thooti	Person who guards the village and implements the orders of the council of elders pertaining to law and order. At the time of British he was also the security guard to the village.
Talwara	An individual who acted as personal attendant to the *Patel* and *Shanubog* (assistant to the *Patel*) and also as a messenger. He was to assist the *Shanubog* in tax collection.

Informal but Recognised Social Institution

The farmers in the tank command respect for the decisions of the informal institution in order to achieve efficiency in water use and best returns from the crops grown with the support from the tank water. The informal institution comprises a council of elders representing all the seven villages, who make decisions on both, the tank and the overall tank water management. There is no evidence as to when and how this informal system for water management started, however a careful examination of the system reveals that it must have started from the time of the *Jodidars*.

Catchment Area

The catchment area of the Mudiyanur tank comprises about 993 ha dryland tract of Mudiyanur (869 ha), Bestrahalli (44 ha), Chikkanahalli and Jigani Bommasandra (150 ha) villages. According to the available secondary data, the maximum elevation difference between the highest and lowest point is about 52 degrees along longitude and it is about 34 degrees along the latitude of the watershed. The average slope of the watershed is sometimes more than 6.5 per cent and hence the soil erosion in the watershed area along the fringes is high. The Mudiyanur catchment area indicates a slope of about 3.5 per cent. Table 6.2 shows the details of the tank.

Table 6.2

Details of the Tank

Sr. No	*Details*	*Units*
1.	Catchment region of the tank	923 ha
2.	Tank area (water spread area)	96 ha
3.	Command area on record	103.6 ha
4.	Command area on cultivation	185 ha
5.	Bund length	990 m
6.	Bund top width	12 m
7.	Bund height	5.8 m
8.	Maximum water level	811.5
9.	No. of sluice	2
10.	Location of sluice	0.09 and 0.9 km
11.	Sluice type	Plug and rod
12.	Sluice sill level	809.19
13.	Number of waste weir	1
14.	Waste weir type	Sloping apron
15.	Waste weir length	30 m
16.	Tank capacity	1.11 H/mcm

Mudiyanur tank is a part of Uttanur watershed. This watershed contains 12 tanks given below:

1. Uthanur, 2. Mudiyanur tank ayacut, 3. Gujjanalli, 4. Malagalamma, 5. Thaverae, 6. Mandigal, 7. Gondalli, 8. Konanahalli, 9. Muruhanakunnu, 10. Bisnalli, 11. Sathanur, 12. Renugunda. The watershed under Mudiyanur tank occupies about 15 per cent of the Uthanur watershed. The villages in the Mudiyanur catchment region include Mudiyanur, Jigani Bommasandra, Bestrahalli, Chikkanahalli and Gujjanalli.

Reasons for Reduced Water in the Tank

There are several reasons for the reduced water flow into the tank. A variety of factors have contributed to the reduced filling of the tank. Based on the opinions of the farmers these reasons include:

a) Reduction in the average annual rainfall in the area over the last few years,

b) Land development activities by the farmers in the catchment area,

c) Plantation/aforestation activity by the forest and horticulture departments in the catchment area,

d) Taking over of *Gomalas* (grazing lands), temple lands and other common lands by the government and their distribution to the poor leading to the reduced water flow to the tank,

e) Implementation of several schemes of the government for the land development activities such as levelling, bunding, construction of farm ponds has further resulted in complete harvesting of surface water, leaving none to flow into the tank, and

f) Accumulation of silt in the tank is yet another important reason for reduced filling of the tank. Construction of a check dam by the Forest Department in the upper reaches has resulted in the reduced inflow into the tank.

Command Area

The command area of Mudiyanur is structured in such a way that it conserves every drop of water harvested into the tank. The structure is so well designed that its efficiency in distribution of water is achieved by a good network of canals and drainage system. Water distribution is through two sluices located on either side of the tank. The canal extending from the sluice runs all along the command area, and is named *rajakaluve*. From this canal, distributaries are drawn both vertically and horizontally. At the lowest points of the command area and at the tail end, arrangements are made to collect the drained water from the fields on either side of the command area.

Divisions within the Command Area

The command area is divided into seven divisions based on the linkage to the distributory. The lands next to the bund receiving water from the right bank canal are identified as *modalagadde*. These are followed by *vemaguntagadde* which are irrigated from another small tank nearby. This is followed by *aiduberala gadde* followed by *baralagadde*; the portion at the lowest point and receiving the seepage water from the tank and the command area. Fields receiving water from the left bank canal are called *vanamu gadde* (garden land) as this portion is elevated compared to the rest of the command area. The entire surplus from all these portions flows into *oddu* or the bund at the lowest point, irrigating a few fields. This collected water is then reused for irrigation purposes. Apart from the canal used in utilising drained water, there are three other points where small bunds called *oddu* are created to tap the surplus water flowing out of the tank along with the seepage from the fields. Water collected at each of these *oddu* is reused for irrigation. Along with these the surplus water flowing from the drylands around the command area and from small ponds in the vicinity of the command area is collected at the *oddu* and used for irrigating the land at the tail-end. Complementing the good network of canals in the command area, the cultivation of crops is also efficiently planned. As the tank is located in a low rainfall region, a catch crop is planned to allow the collection of water to its potential capacity during the monsoon season. This catch crop is partially irrigated and that too for one and a half months in the later stages of the growth. The catch crop is followed by irrigated paddy only during summer months if the tank is holding water to its total potential.

Managing Water Shortages

Managing water shortages in the canal command is one of the important aspects in the overall water management plans. The experience over the last few years has compelled the 'council of elders' to search for alternatives to meet the water shortages faced during summer months even when tank is in at its full capacity. In order to have a strong strategy to mitigate water shortages the management divides the command area into three portions and release water for irrigation to each of the divisions on a rotational basis. This new approach is being put into operation successfully and all farmers in the command area have accepted the decision.

Structure and Functions of the Social Institution for Water Management

The social institution for the management of tank and tank water is the council of elders. This is constituted for the general administration of the seven villages comprising Mudiyanur, Bestrahalli, Chikkanahalli, Jigani Bommasandra, Krishnagiri, Hosur and Kencharanahalli. These villages were traditionally *Jodi* to the *Jodidars* in Mudiyanur village. Another major responsibility of the council of elders is to organise festival for the goddess *'Chowdeshwari'* whose temple is located at the foot of the tank bund. Villagers of these seven villages perform day-to-day *poojas* (worship) to the goddess and also conduct an annual festival.

Structure of the Council of Elders

The council of elders is a two-tier structure. The first tier of the structure is constituted by representatives of each caste group residing in each of command villages and the village functionaries constitute the second tier. The important village functionaries in the second tier include *thooti, neerganti and talwara*. These village functionaries normally belong to the scheduled castes. They are provided with *inam* (free) land in the command area and elsewhere as an honorarium for their services. They also normally act as representatives of their caste at the council of elders. Such representation is not necessarily based on the economic or political position of an individual, but on the capability of the person to command respect from his fellow caste members or his ability in dispersing work to his fellow caste members in a manner beneficial to the caste and to the general welfare of the village.

Services and Payment for the Village Functionaries

At the end of the harvest, the *thooti* gets 6 bundles of paddy, *neerganti* gets 12 and *talwara* 3 from one unit or half acre (*dala*) of land. During a good harvest, *neerganti* may end up in getting 25–35 bags (each bag of 60 kg) of rice and other two get half and quarter of that. In addition, the families of *thooti, neerganti* and *talwara* have been given 16 acres of dry land and 5 acres of irrigated land. The services expected of the respective functionaries are as follows.

Neerganti

- Responsible for ensuring the repairs for the leakages in the sluice, tank and the bund.
- Acts as the information man on water management and takes care of water distribution.
- Plays an important role in conflict resolution.

Thooti

- Responsible for organising village level festivals.
- Organising village meetings.
- Making public announcements in the villages.
- Participating in auspicious (marriages) and inauspicious (death) occasions on behalf of the council of elders.
- Acts as messenger for all announcements relating to the tank and water management.

Talwara

- Attends the council of elders and has watchdog functions.

The council is first constituted at the village level by inviting representatives from each caste residing in the village. Irrespective of its size, one member represents each caste. Each village selects one person to represent the village at the council of elders constituted for performing various activities associated with the temple. Normally this representative will be from the largest caste group and also from that which commands the largest area under cultivation. If there is a division within the caste at the village level due to political or economic rivalry, representation is provided to each factional group —who are asked to re-examine their representation at the council. A representative of a caste can continue as long as he enjoys or commands the support of his fellow caste members in terms of adherence to the decisions arrived at by the council of elders. If the ability is questionable, the one who commands more support may replace him. In the case of village functionaries, the head of the office changes every year on the basis of an understanding arrived at by the families of the scheduled caste. In Mudiyanur there are four households of scheduled caste and each family takes up a different role each year. Only these four families have the right to be the *neerganti* for the Mudiyanur tank, and to claim *inam* land (although this is independent of performance of duty as a *neerganti*). No other family from any other village has the right to be a *neerganti*. Traditionally, the council of elders is headed by one of the *jodidars* from Mudiyanur. Normally, it was the *patel* who heads the council of elders constituted for the temple. Although the post of *patel* was abolished in 1976, the son of the last *patel* is recognised as the head of this council. This continuation is not challenged, although people do recognise the emergence of another capable person. Though this emerging person is not invited officially to the meetings, he is consulted on various issues pertaining to the temple and the tank.

Decisions by the Council of Elders

The major decisions in the case of Mudiyanur tank are related to the release of water for irrigation, attending to repairs required for the infrastructure for effective delivery of water and most importantly the conflict resolution as and when they arise. Decision-making with regard to water and infrastructure management are largely guided by well-established patterns of behaviour. No one is aware of the genesis of these patterns, but they are well-understood and practised by each and every one related to the tank. Any deviation from the pattern is considered as a risk factor and is normally avoided.

Decisions on Crop and Water Management

The decisions on crop and water management are determined by the council of elders. The pattern of water distribution is determined by the elders (predecessors) as established in the past. Meetings and discussions with key informants indicate that the farmers mainly grow different varieties of paddy during the *kharif* season. Direct sowing or broadcasting of seeds is carried out during mid-June as soon as the monsoon sets in. The soil, being clay loam, holds lot of moisture. The tank sluices are not opened until the end of July, at the time of weeding. This saves nearly 30 per cent of the tank water. Earlier if the tank overflowed before June or if the storage reached the optimum level, farmers would transplant the seedlings. Over the years, farmers have experienced water shortages even after the tank filled up in June. Hence they have resorted to direct sowing which suits to the water scarce situation. Moreover, it is very difficult to cultivate other crops in this stretch during rainy season, since the clay loam soils can quickly become waterlogged if there is a sudden outburst of rains. Of late, farmers have started experiencing water shortages for growing paddy in post rainy season or summer, even with the tank overflowing in the month of December. Consequently, better wisdom prevailed on the farmers. From recent past, the council of elders arrived at a conclusion to irrigate only part of the command area (80 acres out of the total 240 acres). The command area was divided into three divisions of 80 acres and in the first year, the first division was irrigated during summer. In the coming year the second division will be taken up for irrigation and in the third year the last division will be taken up for cultivation.

Decisions on Cropping Pattern

The decisions on the cropping pattern are discussed in the council of elders and the cultivar to be grown is decided. The major crop grown in the tank command is paddy. All the command farmers cultivate paddy in both the seasons. The difference between the two cultivating seasons is the cultivar choice which mainly depends on the availability of water for the crop. Usually during the rainy season (monsoon) the local paddy cultivar called *'pedda byra voddalu'* is cultivated. The choice for this cultivar by the command farmers arrives based on the proven ability of the cultivar to come up well in an unpuddled soil bed with not much preparation. This cultivar also has proven abilities to withstand shortage of water or drought conditions. In the second season (summer) there is no restriction on the cultivar to be grown and the individual farmers can decide their own choice of cultivar as the council of elders honors the preferences of the farmers in this regard. However, the preference is for short-term varieties as they reduce the risk of shortages at the end of the season (terminal drought) and the chances of possible losses to the crop by stray cattle grazing, if the duration is beyond the cropping season.

Decisions on the Release of Water

The important decisions concerning water management are made by the council of elders who follow the established patterns on release of water and the consequent practises to be followed in each season by the fellow farmers in the command area. Accordingly, during the rainy season no farmer, irrespective of the level of water stored in the tank, is allowed to draw water for raising the seedbed or puddling the field. Water is not released for irrigation purposes till the commencement of *magge* rainfall. By this time the crop sown would have reached reproductive stage and irrigation will definitely boost the yield. The water will be released for irrigation after the consultative meeting of the council of elders in the first week and last week of December during which the council of elders with the assistance of the *neergantis* and other farmers estimate whether water accumulated in the tank is sufficient for a crop or not. To estimate the sufficiency, there are indicators, which are proven to be practical over the years. According to these indicators, if the water level in the tank is upto the uppermost division marked on the exclusive stone pillar erected for the purpose of measuring the depth of water which is located closer to the sluice, or if the water level touches the horizontally laid stone slabs over the sluices then it is estimated as sufficient to raise a crop of paddy. If the level is far below the set pattern then no crop is raised and water will not be released for irrigation. If the level of water is just few feet lower than the marks for the full tank, then the opinion of the *neerganti* is sought. If the *neerganti* expresses his confidence on the water stored in the tank to help in raising a crop, then the council decides to release water for irrigation. Based on these estimations, the dates for the release of water are decided. Normally, it is the *neerganti* who will advise the council of elders on the exact day for release of water. This suggestion of *neerganti* is essential, as he has to attend to the repairs and maintenance of canals before the release of water.

According to the *neerganti*, distribution of water in summer has to commence in the evening and has to be completed in the night. This will prevent intervention of the landowners and also evaporation. If landowners are allowed to intervene, their demand for storing water has no limit; accommodation of such demands will affect the supply of water to the fields. Based on the stage of the crop, water has to be distributed. If it is in the growing stage, little irrigation is required. On the other hand, at the reproductive stage a good irrigation is essential. Once the grain formation is completed earhead formation has completed, the field can once again be less irrigated.

Decisions on Infrastructure Management

Basic infrastructure for the delivery of water in the command area is regarded as the property to be managed by the council of elders and the rest, including sluices, bunds and other structures in the catchments are not regarded as the properties to be managed by the council. With such options the council of elders seeks the opinion of *neerganti* in the maintenance and management of these properties. Based on the suggestions of *neergantis,* the council advises the farmers to attend to the repairs of the tank infrastructures. Structures that require repairs are the sluices and the delivery canals. As the sluices are built of stone pillars, no repairs are required. But to tap every drop of water in the tank, silt accumulated around the sluice has to be removed. Normally this will be accomplished by the concerned *neerganti.* If the canal has to be drawn from the point where water is accumulated within the bed, then the farmers receiving water from considered sluice will be summoned to make a canal. Farmers attend to the major canals and distributaries on a regular basis. Works in these canals are attended to only if the *neerganti* summons all the farmers served by the canal. Farmers are summoned for two kinds of activities. Firstly, to remove weeds and desilt the canals regularly in each season prior to the release of water, secondly, to attend to major repairs like breach in the canal or strengthening the canal bund with additional reinforcement, etc. To attend to regular weeding and desilting during each season, through informal consultations, the *neerganti* collects the opinion of several farmers, especially those who hold land on a large scale, and fixes the day for weeding and desilting. Normally, the day fixed will be a Monday and the work is scheduled to start during the morning hours. All canals are weeded out and desilted in each season from top to bottom with the collective participation of all the farmers served by that canal. If the work remains incomplete, farmers could be summoned on the next day or on the following Monday. The distributaries have to be desilted by the individual farmers served by it. Normally, the length of the distributary passing next to one's field is desilted. A person failing to participate in the desilting and weeding activities without prior permission is liable for punishment. Punishments are decided by the council of elders based on the complaint filed by the *neerganti* identified for the respective canal. Normally, farmers are let off with verbal sanctions and with a warning not to repeat such actions. In fact, the very act of being questioned by a *neerganti,* who hails normally from a lower caste, is considered a sanction. Farmers from upper castes generally avoid getting into situations which will entail questioning by the *neerganti*. If the matter does come to the council of elders the representative of that village or the caste leader may be asked to reprimand the actions of the particular farmer. The council of elders has the right to stop the supply of water to the fields of the offender. In some cases, release of water may be delayed in order to exact a change in behaviour, although the extreme step of cutting off irrigation supply entirely is not generally exercised.

Role of Caste and Class in the Council of Elders

The management decision-making process has clearly evolved along the lines of caste hierarchy. The *jodidars* are Brahmins, and they continue to head the council. This leadership has evolved out of both the prominence of the temple and the ownership of land. However, with the gradual transfer of ownership of land to other caste groups, the process of consultation with all landowners is reducing the domination of the Brahmins. This process of consultation is once again based on the size of the caste and its importance in carrying out certain activities of the temple, especially at the time of the annual festival. While reckoning caste as the basis for representation, it is not proportional representation, as each and every caste in the village is represented irrespective of its size. Even if there is only one family belonging to a caste, it will be represented in the council. This obviously benefits those castes which are small in number and who cannot muster support in terms of number of votes. At the village level, the rich are generally selected as leaders of a caste and as members of the council of elders. This may not happen to all situations. It is expected that the leader should have the abilities to avail services from the government, elected leaders and officials at various levels; get the work done, assist people in distress, get justice in conflict resolution, and above all, discuss and convince others about his view point. Villagers of Bestrahalli of the *vokkaliga* caste select their leaders not only based on wealth but also on the overall acceptance by other caste groups. With the same criteria, the *thogata* caste group also selects their leader.

The Link between Wealth and Leadership

According to key informants, the person selected has to be wealthy, as his absence from work has to be replaced by a paid labourer. Expenses to meet the officials and politicians have to be met by him. Above all, he has to host these officials and politicians if they visit the village—and he needs to have the resources to do this. The selected leader does not attend the meetings alone. Those who can convince others in the caste, especially the potential opposition within the caste, will accompany him. Villagers do not select the rich alone as their leader and also the decisions of the council will be in consensus with all the people and not just favouring the rich. Decisions of the council of elders are always implemented by the scheduled caste. Traditionally, the three positions of *thooti, neerganti* and *talwara* are reserved for the scheduled caste unless totally rejected by the scheduled caste families. These people are supposed to be the eyes of the council of elders. As noted earlier, each village functionary and their respective works are clearly defined. Payments are also very well proscribed.

Decision-making Processes and Conflict Resolution

According to one of the leaders of the council, decision-making (on punishments) is difficult, as you live with the same person in the same village and participate along with that person in all activities and share joy and sorrow equally with him. Normally, when a decision has to be reached on an issue, the members of the council informally explore how the concerned issue is understood, and what villagers expect. If the opinion of the villagers is against the legal domain outside the village, the leaders explain the consequences in an informal way. Further, informally, each member will explore all possible alternatives available and the potential response to each alternative before pronouncing a decision.

Conflict Resolution

The feeling that judgement should be helpful to a plaintiff to reconcile the friendship or relation in future with the person or parties against whom he has complained, acts as the guiding force in resolving conflicts at the council. However, one essential step that is invariably followed is to first consult the relatives of the parties involved in the conflict and also those who help them in carrying out economic, agricultural and religious activities. In fact, these parties are consulted before meeting the party involved in the conflict. This consultation is normally carried out to understand the point of view of the relatives and others who provide support to the party and also with an assumption that the party involved must have shared its views on the issue with these people. After a series of such consultations, the members of the council arrive at a most feasible and plausible decision. Punishments are decided in a way that it harms the parties or culprits in least terms. The decision should cause least damage to his/her status and reputation. According to the council of elders, the decision should never lead to caste clashes. It should not create hatred among the parties involved and it should keep avenues open for reconcilements between the two parties at any time. The relatives of both the conflicting parties educate and prepare the parties involved to accept the decision to be pronounced by the council of elders. On the day of verdict, both the parties are allowed to express their views, feelings and emotions, but the parties involved in informal decisions will be acting as guards in not allowing the situations to go beyond certain level. This way of resolving a conflict by involving relatives and those who matter the most and not the party concerned protects the position of the leaders in the council. Such guarding is considered essential for the leader to be perceived as impartial. This methodology of resolving conflicts has been helpful over the years. The absence of monetary punishment for those who fail to contribute labour for maintenance of canals and to those who water their fields against the rules, speaks in favour of the effectiveness of compromise and reconciliation. By involving individuals beyond the conflicting parties, a level of transparency is achieved.

Weakness in the System

Whilst the council of elders system of decision-making has operated effectively for many generations, it has a number of inherent weaknesses, especially when viewed in a modern perspective. These are both at the social, and more technical (water management) level.

Changing Social Norms

As noted earlier, the village functionaries are broadly positive about the traditional system of tank management, noting that it provides them both with respect (even if this is thought to be less than earlier) and a secure livelihood. Nevertheless, as a system based on the maintenance of caste hierarchy and status quo, the validity or authority of the traditional council of elders in modern, democratic times may be questioned. Whatever respect is afforded to women, the decision-making system does not give them a voice; whatever respect and remuneration given to the village functionaries, they are not expected to step beyond their allotted caste-defined role. Is this equality or democracy? The council of elders themselves note that there are many changes in the villages, and that these changes pose a major challenge to them. According to them, these changes increase the number of absentee landlords; lead to the emergence of new leaders; and bring in ability among the villagers to develop contact with outsiders. Changes brought in by the government have also reduced their authority.

Increase in the Number of Absentee Landlords

Changes in the families of *jodidars* in terms of achievements in education and consequent employment outside the village have increased the number of absentee landlords. According to one estimate, such absentee landlords command three-fourths of the command area. The chances of their return to the village are remote. They take less pain to maximise the returns from the fields that they own in the command area. Thus, they fail to contribute to development of the tank. The tenants also do not cooperate with the council, as their employment for the next crop season as tenants is not assured. Many times the agreement between the tenant and the landlord is not finalised by the time maintenance activities have to be initiated. Hence, as complained by the *neergantis* it is difficult to summon farmers to maintain the infrastructure in the command area.

Emergence of New Leaders

The youth, especially from the traditionally marginalised castes, are able to emerge as leaders with good access to education and employment opportunities and political backing. New leaders and the leaders of the council of elders fight for recognition. The process and style of functioning of the council of elders are different from the functioning of other systems that generate new leaders.

According to the council of elders, they work informally and the process involved in their system of administration is time consuming. Comparatively, the new leaders want to be formal and approach the institutions outside the village for speedy settlements. The council of elders feel that this may cost more in terms of finance and is potentially manipulative. Consequently, the conflict in the village will assume greater proportions and last longer.

Ability to Contact Outsiders

Politicians outside the village often become involved in playing power games within the village. With education and increased awareness, villagers (particularly the youth) have become aware of how to approach higher officials and claim benefits from the government.

Changes Introduced by Outsiders

The council of elders tends to be hostile towards attempts made by outsiders, particularly the government, to promote participatory processes of decision-making. They see the creation of user groups as threatening for the following reasons:

- Benefits can no longer be dictated by the council of elders, but must be shared equally.
- The leaders have become answerable to the questions of the 'lowest' in the social hierarchy.

The leaders of the *jodidars* are particularly alert to analyse the emergence of user groups. They note that the *jodidars* no longer enjoy total control over all the activities within the seven villages, having lost control over several institutions and activities. The first to go away from their hands was the school. Then it was the cooperative society for milk, followed by the *gram panchayat*. Now the attempts to form user groups for watersheds and tanks are seen as attempts to snatch away the institutions from their hold. Hence, attempts to form a user group are being prevented by the council of elders by ordering the people not to contribute any amount or to participate in those activities.

Water Management Challenges in Current Times

Catchments or Watershed

Even with an understanding that the system is very well related to the status of the catchments, the institutional structure is not able to address the issues of the catchment effectively. The council of elders has not taken up the issue of changes in the land use pattern or the diversion of runoff from the tank. Though admitting the tank is the lifeline of the poor villagers around the tank, the present system has totally failed to check silting of the tank.

Drinking Water

Ground water depletion has lead to drinking water scarcity in the study area. Though the government agencies have taken appropriate steps to provide drinking water facilities, the problem of declining ground water and consequent impact on its availability are addressed neither by these agencies nor by the council of elders.

Control Over Groundwater Exploitation

The institutional structure for tank management has not been able to address ground water exploitation, in spite of awareness of the impact of bore-wells on the ground water depletion. The institutional structure has not checked drilling of borewells close to the tank. They have no norms to control pumping of water from a well located in the command area, which is taken to a far-off place. The management of Mudiyanur tank provides an example of a still functioning traditional system of decision-making, a system that was common in the area in earlier times but which is found less these days. Whilst the system has flaws—and may be particularly criticised for the lack of participation of women and the manner in which roles are caste determined—it can nevertheless provide interesting lessons. Building and enhancing traditional systems of management are often more effective than introducing completely new mechanisms. Understanding the traditional system provides an insight into the tank as a natural resource that is the focus of a social system. Secondly, it provides insight into the methodology of conflict resolution.

Sharing of Resources by more than one Community

The common resource is shared by seven villages. This provides an example of the management of a resource used by multiple communities. The formation of the council of elders and the power shared by each council at each level provides insights in developing a system for the management of a resource owned or used by many communities. The council at each level being responsible to its members is also an example of decentralised power system.

Adaptation to Short Supplies

The way in which water shortages are shared not only in Mudiyanur, but also in the tanks of Mandikal, Kothamangala, etc., provides insight on how farmers tailor their paddy cultivation to short and uncertain water supplies. As has been seen, water use is judiciously regulated, paddy varieties chosen carefully, and the method of cultivation adapted to low water availability. An additional important point is that shortages in water supplies are shared in a manner that is accepted by the community members in an equitable way.

Conflict Resolution

Perhaps the most important learning from the example of Mudiyanur tank is in terms of understanding traditional approaches to conflict resolution. The important aspects of this are: (a) the parties have to live together for the rest of their lives, (b) escalation of conflict is strenuously avoided, (c) persons on whom the parties are emotionally or economically dependent are involved in the resolution process, and (d) penalties are chosen in a manner that avoids loss of face.

The underlying principles in conflict resolution is to:

(1) Consider the parties in conflict as people who will live the rest of their lives in the same village. Hence, resolutions are framed to allow both parties to develop rapprochement in an honorable manner. The council of elders even point to cases in which parties once in conflict have discarded their differences to the extent of developing martial relations between their families. If the conflicting parties own land next to one another, conflict resolution demands that they are able to tolerate intrusion of animals into each other's fields, to cooperate with one another in managing surface runoff, and in other agricultural activities. The elders report that the failure to manage surface canals in the catchments these days is due to the lack of cooperation between neighbouring farmers.

(2) Another principle in resolving the conflict is to avoid it escalating into conflict between kin or caste groups. Leaders fear that such

escalations are not only dangerous to the unity of the village, but also difficult to resolve, as positions become increasingly intransigent at each level of escalation. This is avoided by involving the kin of the parties as negotiators at the first level, so that they do not act as agents for escalation, but rather as agents who can check emotions, and have a personal responsibility for resolving the conflict. If the parties are from different castes it is necessary to involve the leaders of each caste in the resolution process.

(3) Involving friends, family and those on whom the parties are otherwise dependent in the resolution process. It is the firm belief of the elders at the council that parties involved in a conflict will be charged with emotions. Such emotions are strengthened and sustained by family members. To avoid the escalation of such emotions, it is essential for the leaders to dampen them at the initial stages. This is possible only when the leaders are aware of the details of all concerned in terms of their economic, social and emotional dependencies. Leaders feel that by summoning individuals who provide support to the parties concerned, and involving them in resolving the issue, the conflict is not allowed to spread. Those who would be the potential agents in spreading the conflict are instead charged with the responsibility of resolving it.

(4) Determining the penalties which avoid loss of face. To ensure better and continued management of the canals, conflicts should be resolved in such a way that the event is forgotten as quickly as possible. This is only possible when the punishments awarded do not affect the position of a person in the village, and the conflict can be quickly forgotten. Thus it must be done without bitterness, and without the penalty or the cost resting in the memory. In summary, the manner of conflict resolution that has evolved in Mudiyanur—and no doubt other traditional systems—is one from which there is considerable potential to learn. It may be argued that when designing new institutional mechanisms for tank management, building on the positive aspects of traditional systems is likely to result in greater institutional sustainability.

References

Government of Karnataka (2002). *Report of High Power Committee For Redressal of Regional Imbalances.*

Intercooperation in India (2005). "Tradition Meeting Modernity: A Case Study on the Management of Mudiyanur Tank, Kolar District", *Karnataka Working Paper* 2, Intercooperation Delegation, Hyderabad, India. p. 33.

Madangopal (2002). "Introductory Speech" to Consultative group on legal Aspects for Community Based Tank Rehabilitation Project.

Reddy, S.T. (1997). *Impact of Commercial Crops on Farmer Managed Irrigation Systems*. PhD, Thesis submitted to Dr. Ambedkar Open University, Hyderabad.

Reddy, S.T. and K.M. Anantharamaiah (2002). *Institutional Approach to Community Based Tank Rehabilitation in Karnataka.* Submitted to Jala Samvardane Yojana Sangha, Government of Karnataka, Bangalore.

Reddy, S.T. (1991). *Forfeited Treasure.* Bangalore: Prarambha.

Sengupta, N. (1993). *User Friendly Irrigation Systems.* New Delhi: Sage Publications.

World Commission of Dams (2001). *Report on India.*

7

Dynamics and Sustainability of Common Pool Resources

A Case Study on the Management of Sacred Groves in Kerala

SUNIL D. SANTHA and K.S. RATHEESHKUMAR

Introduction

The forestry sector is an important source of food, employment, revenue and sustenance for many traditional communities in the developing world. Today, forest resources are said to be in a crisis of overexploitation, species extinction and resource degradation. Various factors such as indiscriminate and destructive felling practices, encroachment of forest lands for the cultivation of cash crops, mono-cropping etc., are found to be some of the major factors that have led to the depletion of forest resources. Another important factor that has been pointed out by researchers is the failure of effective institutions in managing the forest resources. Poor participation of local communities in the management of forests is considered to be one of the main causes for the ineffectiveness of the present day resource management institutions.

While the crises in protected and reserved forest management have received much attention of researchers, this has not been the same with other kinds of forest resources such as sacred groves. Sacred groves are plots of natural vegetation surviving amidst man-modified landscapes; and owe their preservation to the perceived importance to some form of divinity (Jayarajan, 2004). These segments of land contain trees and other forms of life and geographical features that are delimited and protected by human societies believing that preserving such a patch of vegetation in a relatively undisturbed state is necessary for expressing one's relation to the divine or to nature (Chandran *et al.*, 1997). These resource systems also have the characteristics of common pool resources (CPRs) and explain the dynamics of natural resource management (NRM) in a better fashion. However, these productive ecosystems are under pressure from the needs of a growing population, pollution and habitat degradation.

In Kerala, sacred groves are spread across the West Coast to the eastern highlands. They are known by different local names such as *Kavu, Sarpakkavu, Nagam, Mundya, Vallikkettu* etc., (Jayarajan, 2004). The history of sacred groves in Kerala can be traced back to the hunting-gatherer societies, who attributed sacred values to segments of forests within their territories (Chandran *et al.*, 1997). Apart from the divine beliefs, people also maintained sacred groves so as to preserve the medicinal plants in their locality (Ramachandran and Mohanan, 1991). As debates across the world emphasise on user group participation in the management of natural resources such as forests, this study examines the dynamics and complexities involved in the management of forest resources such as sacred groves. Considering the case of sacred groves management in Thazhava village of Kerala, this paper demonstrates that resource management institutions evolve and are continuously shaped by a process through which people consciously and unconsciously draw on existing social and cultural arrangements in response to the changing situations. This paper also examines those factors that contribute to the sustainable management of the sacred groves amidst the inherent dynamics of resource management.

Theoretical Background

Approaches to understand human behaviour, resource use and environmental degradation in the 1960s were predominantly guided by the neo-Malthusian viewpoints that overpopulation and the limited availability of natural resources would threaten economic growth. The publication of Garrett Hardin's *The Tragedy of the Commons* in 1968 marked a significant step towards understanding the problems pertaining to the management of the commons. Looking at problems of population growth, pollution and environmental degradation, Hardin argued that the commons would be subjected to rapid and severe resource degradation due to its very nature of being managed by a collective group.

Citing the example of a 'pasture open to all', Hardin attempted to show that each herdsman would try to keep as many cattle as possible on this pasture for his own benefit. Overgrazing of the resource and its degradation will be the result, when every herdsman keeps on adding cattle to his respective herd. Hardin visualised that, 'ruin is the destination toward which all men rush, each pursuing his own best interest in a society that believes in the freedom of the commons. Freedom in a commons brings ruin to all' (Hardin, 1968: 1244-1245) Hardin and his followers argued that in situations where exclusion from receiving the benefits becomes difficult, individuals tend to free ride on the efforts of others rather than contributing to it. An alternative that was suggested by them to prevent the over-use of natural resources was instituting a system of private property or state property (external control). Though initially popular among the policymakers, resource managers and nation-states, this approach began to receive much criticism in the later years.

Case studies from across the world proved contrary to the belief that natural resources could be effectively managed as a common property through a collective. Hardin was criticised for conceptually mistaking common property resources as synonymous with open-access resources. Scholars from different disciplines have examined the different rules governing resource use, the features of successful institutions, and the conditions conducive to the emergence of collective action. Their work has been collectively referred to as the collective action school or as 'CPR theory' (Steins *et al.*, 2000; Husain and Bhattacharya, 2004). Their studies provide an in-depth analysis of local, community based efforts to manage and govern CPRs. Among these, the work by Elinor Ostrom (1990) is considered to be the most influential and significant. Other major contributors to the collective action school are Wade (1988), Berkes (1989), Bromley (1992), and Baland and Platteau (1996). These studies argued that individuals dependent on a common property could work together in framing rules to regulate benefits produced by the resource, which could facilitate the sustainable management of the resources.

Ostrom relied on information from detailed cases studied by scholars across the world. She examined many field experiences of success and failures (14 cases) across a wide spectrum of countries and sectors such as management of pastures, forests, irrigation tanks, fisheries etc., with a view to identify conditions that are conducive to effective local level management of CPRs. According to Ostrom, the problems of CPRs as visualised by Hardin and his followers, cannot be applied to an entire set of CPR situations. She criticises the approach of both privatisation and nationalisation (centralisation) of commons. Her effort was to bring out a theoretical and empirical alternative to manage the commons. Based on her analysis, Ostrom has formulated seven 'design principles'. A design principle is considered to be 'an essential element or condition that helps to account for the success of these institutions in sustaining the CPRs and gaining the compliance of generation after generation of appropriators to the rules in use' (Ostrom, 1990: 90). These design principles characterise all the robust CPR institutions. She adds one more principle as the eighth principle that is suited for larger, complex cases. The design principles are listed in Table 7.1.

Table 7.1

Design Principles as Proposed by Elinor Ostrom

1. **Clearly defined boundaries**

   Individuals or households who have rights to withdraw resource units from the CPR must be clearly defined, as must the boundaries of the CPR itself.

2. **Congruence between appropriation and provision rules and local conditions**

   Appropriation rules restricting time, place, technology, and/or quantity of resource units are related to local conditions and to provision rules requiring labour, material and/or money.

3. **Collective-choice arrangements**

   Most individuals affected by the operational rules can participate in modifying the operational rules.

4. **Monitoring**

   Monitors, who actively audit CPR conditions and appropriator behaviour, are accountable to the appropriators or are the appropriators.

5. **Graduated sanctions**

   Appropriators who violate operational rules are likely to be assessed graduated sanctions (depending on the seriousness and context of the offence) by other appropriators, by officials accountable to these appropriators, or by both.

6. **Conflict-resolution mechanisms**

   Appropriators and their officials have rapid access to low-cost arenas to resolve conflicts among appropriators or between appropriators and officials.

7. **Minimal recognition of rights to organise**

   The rights of appropriators to devise their own institutions are not challenged by external governmental authorities.

8. **Nested enterprise for CPRs that are part of larger systems**

   Appropriation, provision, monitoring, enforcement, conflict resolution and governance activities are organised in multiple layers of nested enterprises.

Source : Ostrom, 1990: 90.

Ostrom asserts that these design principles play an important role as they could affect incentives in such a way that resource users will be willing to conform to operational rules devised in such systems, to monitor each other's actions with respect to these rules and to replicate the CPR institutions across generations. The collective action models, Ostrom's design principles in particular, continue to provide a significant lead in addressing issues pertaining to the tragedy of the commons, as well as in evolving new institutional models for resource management. An enriched understanding of social, ecological, economic and political dimensions of natural resource management has been generated. Policymakers, non-governmental organisations (NGOs) and multilateral aid agencies have adopted these models, as tools for crafting institutions for co-managing natural resources. However, of late, these approaches have also opened up serious debates among social scientists on issues relating to NRM. Some of the key issues debated are discussed below.

It is often debated that the collective action approaches provide less importance to contextual factors that influence NRM. Contextual factors include the social, cultural, economic, political, technological and institutional features of user groups. Changes in culture and social structure, issues of internal differentiations such as plurality of beliefs, norms, equity and power relationships between resource users and their implications on resource management are given little emphasis by the collective action models. Examples of contextual factors also include characteristics of the local ecological system, cultural characteristics of the user groups, extent of development of transport facilities linking the community to nearby markets; or changes in government policy relating to resource use, changes in world market prices, developments in labour market outside the community etc. Critics have suggested that the interplay between these contextual factors could result in various complexities and uncertainties in resource use situations, which need to be examined and studied. Changes in the physical environment, economy, social structures and cultural practices themselves could become sources of uncertainty in resource management. Critics like Long (2001) and Steins (2001) point out that in such a situation, there is a vital need to recognise the stakeholders' construction of resource management.

The prominent collective action models consider homogeneity among members as an advantage for crafting effective resource management institutions. Thus, a community in a natural resource management setting is often seen as entities that are bounded, homogeneous, local and identified to a particular user group. The social differences and the diverse, (sometimes contradicting) concerns of resource users are not given any recognition in these processes. Collective action approaches in NRM have to be strengthened in this regard by looking at how users interact and manage resources in a heterogeneous setting.

Ostrom defines institution as, 'the sets of working rules that are used to determine who is eligible to make decisions in some arena, what actions are allowed or constrained, what aggregation rules will be used, what procedures must be followed, what information must or must not be provided, and what payoffs will be assigned to individuals dependent on their actions' (Ostrom, 1990: 51). As per this conceptualisation, a common assumption of collective action approaches is that institutions could be crafted for specific purposes. Institutions are considered as 'rules of the game' designed to achieve the goals of resource management. However, this conceptualisation of institution has received much criticism. Some prominent critics like Cleaver (1998, 2001) and Mehta *et al.* (2001) have opined that institutions need not always be functional or purposive, and can evolve even through unconscious interactions/practices of actors. Collective action is thus seen as an institutionalised set of procedures to regulate human behaviour. On the other hand, Giddens (1984) has shown that institutions not only restrain, but also play a crucial role in enabling and empowering human communities. Individual action is both enabled and constrained by social structure. Institutions evolve and are continuously shaped by a process through which people consciously and unconsciously draw on existing social and cultural arrangements in response to the changing situations.

For example, the key aspect that is often given importance while crafting resource management institutions is strengthening its regulative function. Jentoft *et al.* (1998) and Jentoft (2004) state that institutions need not only have regulative features, but also could have normative and cognitive structures. Institutions are thus inextricably linked with people's cultures, belief and life world. Jentoft also feels that collective action approaches should also provide sufficient scope or accommodate to explain the cognitive and normative functions of institutions. This would help in understanding the dynamics and complex nature of resource management institutions and also explain issues of equity and power relationships in natural resource management.

Collective action approaches based on the design principles often give the impression that culture and social structure will remain static. Local conditions of resource management could change due to both internal and external factors and are never static. Social relationships between various stakeholders, and the components of culture, which guide the use of resource, could change with respect to time and space factors. For example, the assumption that property rights will remain static and that suitable institutions could be crafted to manage these resources is widely contested. Critics point out that such assumption could work out well among traditional self-contained societies, but not necessarily among complex societies where there is a mix of different property regimes. With change in resource conditions or technologies used for appropriation, what was once a private property could either become a state property or be collectively managed by a group of primary stakeholders. Moreover, different types of property regimes may exist at the same time. Further research is required to explain why property regimes change over time and under what conditions will these regimes work or fail. For this, the cultural embeddedness of user groups with respect to their resources, economy and society needs to be addressed.

It is in the context of the above debates that researchers across many parts of the world are exploring alternative explanations and meanings of institutions, collective action and resource management. While the earlier approaches focused on local, homogeneous communities, emerging views attempt to understand communities as heterogeneous entities, located in multiple sites with varying knowledge systems and choices in NRM. Such an approach also helps to understand the conflicting interests, perceptions and livelihoods that exist within or between communities. The collective action approaches in the background of the above-mentioned debates form the theoretical base for the present study. Taking the case of sacred groves in Kerala, this study looks at these issues in-depth.

Methodology and Field Setting

This study is primarily qualitative in nature. Ethnographic field work was carried out in Thazhava village of Kollam district in Kerala. As per local understanding, there are 33 sacred groves in the village. In the study region, the sacred groves are popularly known as *Kavu*. It was observed that a particular sacred grove is dedicated to a specific God, Goddess or a serpent. Primary data was collected through in-depth interviews, group interviews, and observations. The different stakeholders of the sacred groves were interviewed. They include the family who legally inherit the sacred groves, the members of the public who monitor the day-to-day wellbeing of the sacred groves, and the *vedar* tribesmen who have the sole right to cleanse and maintain the sacred groves. The principle of saturation was applied in deciding the sample size, and purposive snowball sampling was the methodology. This particular sampling strategy was used as: (i) the user groups were wide and varied, and (ii) to gather relevant information based on the research questions and emerging gaps at the field. Qualitative techniques of data analysis such as coding, pattern coding, memoing, summary building etc., were used.

Continuity and Change in the Management of Sacred Groves

The findings discussed below are based on preliminary analysis of the data collected. A common pattern that has emerged is that the management of sacred groves is a very dynamic and cultural-specific process. Resource nurturing, conservation and equitable distribution can be identified as the three essential characteristics of sustainable resource use. The first two functions of nurturing and conservation are directly verifiable in the case of sacred groves. On the other hand, the third function of distribution is significantly vested with the divine blessings attributed to the sacred groves. The dynamic nature of resource management can be observed in all the three functions. To be specific, these dynamics can be observed in terms of: (i) variations in resource use and ownership, (ii) the disappearance of traditions and the emergence of new rituals, (iii) latent manifestation and decline of resource use norms, and (iv) decline and re-emergence of institutions in resource management. These aspects are discussed in detail below.

Malhotra *et al.* (2001) has neatly briefed the common property characteristics of sacred groves. Our present study shows that in contrast to the State managed forest resources, the sacred groves are inherited by private family members. However, apart from the family members, there are two other major stakeholders involved in the management of the groves. They are namely, the believers from the public and the *Vedar* tribesmen. The *Vedars* are vested with the sole right to cleanse and maintain the groves. The presence of these two stakeholders changes the whole resource use and ownership characteristics of the groves from that of an exclusive private property to a common property. However, it was observed that based on changes in contextual factors, the property rights shifted from that of a CPR to a private property or vice-versa. These aspects could be elaborated as follows.

Though the legal property rights are vested with a particular family, the utilisation of services is not restricted to them. In the study area, most of the sacred groves are owned by the Nair community. As mentioned above, the usufruct rights are extended to the believers and the *Vedars*. Any other trespassers are fined and sanctions are imposed upon them. The local community is capable of distinguishing the believers from the non-believers. However, the sect of believers hails not only from the Hindu community but also from the Muslim community of the locality. On the other hand, the *Vedars* are tribal folk who reside more than 30 kilometres away from the village under study. These *Vedars* have the exclusive right to enter into the sacred groves at any point of time, cleanse the groves, and also extract canes, honey and other essential non-timber forest products for their livelihood.

There are several norms and associated beliefs that govern the management of the sacred groves. One of the most prominent norms is that the *Vedar* community is only allowed to enter the grove and collect non-timber forest produces from the grove. Anybody else entering the grove with the purpose of resource extraction is warned by the local community. Such norms are also supported by belief systems and superstitions. For example, it is widely believed that hampering the *sarpa kavu* (sacred groves whose main deity is the serpent) will invite the curse of serpents that will be carried over generations. Local communities also substantiate these beliefs with the narration of true incidents that happened in the village. There are also other norms related to the extraction of resources. Even the *Vedars* are only allowed to cut the dry woods. There are instances where the local communities have prevented the *Vedars* from entering the sacred grove as they violated this norm. There is yet another instance which demonstrates that belief systems restrict encroachments and theft of resources. There is a sacred grove named *'Bhoota kavu'* (sacred grove of the devil). It is widely believed that a devil resides in this particular grove. People associate the devil's presence with the death of livestock and human beings in the surroundings. As a result, no one visits this grove and neither any ritual nor worship takes place. Nevertheless, this grove has one of the richest biodiversity in the region.

It has to be noted that the day-to-day rituals at the grove (such as lighting the lamp), especially if the deity is a goddess, is carried out by a Brahmin priest or by the head-woman of the family owning the grove. During the annual festival time, the believers from the four zones (*karakkar*)—north, south, east and west—of the village, take up the responsibility of organising the rituals and events. It is interesting to note that the customary rights are given to the *Vedars* based on certain belief systems. According to the local communities, the *Vedars* are supposed to be the descendants of Lord Shiva and Goddess Parvati. The Hindu mythology describes instances were Lord Shiva and Goddess Parvati disguised themselves as hunter-gatherers. The *Vedars* are also part of a hunter-gathering tribe. Moreover, these assumptions are strengthened by the belief system that snakes in the sacred groves run away from the *Vedars* or that the reptiles do not harm them etc. In addition, these belief systems and the granting of usufruct rights are strengthened by certain rituals that the *Vedars* follow. Prior to the commencement of any work, the *Vedars* offer arrack (liquor) and betel leaves to appease the deities in the sacred groves.

However, over the years, believers among the local community have also begun to offer liquor to appease the deities. Earlier, the practice was to brew the liquor locally and offer it to the earth. No one used to consume it. However, these days they buy Indian made foreign liquors from the beverage corporation. Moreover, the believers consume the liquor instead of offering it to the earth. Thus, we can also see that the divine purpose of a ritual has now transformed in to a social evil pushing many people into alcoholism. This trend has now compelled many owners of the sacred grove to seek refined rituals such as *sapthaham* and *sarpabali* as an alternative.

This change, along with many other sociopolitical and economic factors has resulted in the alterations of traditional roles that the various stakeholders used to play. This has also resulted in the exit of traditional actors and the entrance of new ones. For instance, these days, many of the sacred groves do not get the services of the *Vedars*. Most of the *Vedars* have migrated to far-off towns and cities in search of better livelihood. Analysing the history of relations between the state and the *Vedars*, it could be seen that most of the welfare plans sketched out for the *Vedars* were never implemented. Thus, in terms of access to decision-making and resources, occupation, and social status, the *Vedars* were always marginalised. Amidst all the mythological 'blessings and curses', the *Vedars* are left to mend for themselves. In the absence of *Vedars*, the family that owns the sacred grove has begun to invite other communities for the annual maintenance of the grove. Thus, it could be seen that the sacred groves are a dynamic network of institutions and institutional actors.

In a similar vein, the caste differences also result in alterations in the management of the groves. This could be explained through the illustration of a sacred grove managed by a *Nair* family. Every year, a ritual dance '*thullal*' is performed in the sacred grove, where the dancer (oracle) gets into a trance and predicts the desire of the presiding deity. The dancer is usually from another community. Over the years, many people have lost faith in some oracles due to their indiscipline and dishonesty. As a result, some families have prevented these dancers from participating in the ritual dance. This has resulted in conflicts between the different stakeholders, and has often led to the discontinuation of certain rituals.

It has also been observed that formal-informal interfaces (or state-community interfaces) have influenced the governance of sacred groves. The families owning the sacred grove have begun to show a greater interest in the day-to-day management of the groves, expecting financial aid from the State Forest Department. These developments are based on certain initiatives by the Forest Department to provide grant for the management of the groves. However, such processes have resulted in the clear-cut demarcation of the property through common fencing. The rise in population and the scarcity of land resources have also facilitated the territorialisation of sacred groves. These processes again reflect upon the dynamic nature of property regimes. Similarly interventions by the state authorities have resulted in the wading away of certain rituals such as the blood sacrifice of common fowls. The shift in the offering of locally made liquor to Indian-made foreign liquor is yet another instance of such interfaces.

Discussion

By studying the management of sacred groves in Thazhava village of Kerala, this paper proves that resource management institutions evolve and are continuously shaped by a process through which people consciously and unconsciously draw on existing social and cultural arrangements. Existing theories of CPR management often give the impression that culture and social structure will remain static. On the other hand, local conditions of resource management could change due to both internal and external factors and are never static. With change in resource conditions or technologies used for appropriation, what was once a private property could either become a state property or be collectively managed by a group of primary stakeholders. Moreover, different types of property regimes may exist at the same time. In this context, the term 'common property' and the classification of resources require a thorough re-definition.

This paper demonstrates that norms and practices of resource allocation have significant influence on the lives of local communities and in their day-to-day interaction with the sacred groves. However, it has to be understood that these norms cannot be seen in isolation, but are intimately linked to other components of culture, namely the beliefs and value systems, local knowledge, and other norms of conservation. Nevertheless, rule enforcement by local communities could be inherent in one's day-to-day activity rather than conscious interventions, which are often followed in the absence of any enforcing authority. It was found that these norms evolve, modify or disappear due to the influence of contextual factors such as state intervention, capacity of the resource system, socioeconomic opportunities of various actors etc.

It could also be seen that cultural practices have a significant impact on the environment. The variations in the components of culture such as norms, beliefs and rituals affect the use and abuse of natural resources. The contextual factors also influence decisions of who will manage these resources and in what manner. On the other hand, cultural components are in turn shaped by the changes happening to the environment. From the very case of sacred groves, it is evident that societies derive benefits in terms of livelihood, collective action and spiritual existence from the environment. Resource management practices and arrangements vary with time and space. Therefore, crafting local level institutions for the management of natural resources has to take into account the dynamic nature of culture and the various contextual factors influencing it.

This paper has contributed to the present debates in common property theory by taking the case of sacred groves management. It has highlighted the importance of analysing contextual factors in NRM. This also explains further that in such a resource setting, resource management should be seen as a process involving a complex mix of informal-formal institutions. Any external intervention oriented towards resource management, development or welfare should recognise the existence of such institutions at the community level. The inferences of the present study could thus help in strengthening the initiatives of community based forest management in the State.

References

Baland, Jean-Marie and Jean-Philippe Platteau (1996). *Halting Degradation of Natural Resources: Is there a Role for Rural Communities?*. Oxford: Food and Agriculture Organisation and Clarendon Press.

Berkes, Fikret (ed.) (1989). *Common Property Resources: Ecology and Community based Sustainable Development*. London: Belhaven Press.

Bromley, David William, (ed.) (1992). *Making the Commons Work: Theory, Practice, and Policy*. San Francisco: Institute for Contemporary Studies.

Chandran, Subash M.D., J.D. Hughes and M. Gadgil (1997). "Sacred Groves of the Western Ghats of India—Peechi", Proceedings of KFRI Workshop.

Cleaver, Frances (1998). "Moral Ecological Rationality, Institutions and the Management of Communal Resources", Proceedings of the seventh annual conference of the *International Association for the Study of Common Property*, Vancouver, British Columbia, Canada, June 10-14.

———. (2001). "Institutional Bricolage, Conflict and Cooperation in Usangue, Tanzania", *IDS Bulletin* 32(4): 26-35.

Giddens, Anthony (1984). *The Constitution of Society: Outline of the Theory of Structuration*. Cambridge: Polity Press.

Hardin, Garrett (1968). "The Tragedy of the Commons", *Science* 162(3859): 1243-48.

Husain, Zakir and Rabindra Bhattacharya (2004). "Common Pool Resources and Contextual Factors: Evolution of a Fishermen's Cooperative in Calcutta", *Ecological Economics* 50(3-4): 201-217.

Jayarajan, M. (2004). *Sacred Groves of North Malabar*. Thiruvananthapuram: Kerala Research Programme on Local Level Development, Center for Development Studies.

Jentoft, S., B.J. McKay and D.C. Wilson (1998). "Social Theory and Fisheries Co-management", *Marine Policy* 22(4-5): 423-436.

Jentoft, Svein (2004). "Institutions in Fisheries: What they are, What they do and how they Change", *Marine Policy* 28(2): 137-149.

Long, Norman (2001). *Development Sociology: Actor Perspectives*. London and New York: Routledge.

Malhotra, C. Kailash, Yogesh Gokhale, Sudipto Chatterjee, and Sanjeev Srivastava (2001). *Cultural and Ecological Dimensions of Sacred Groves in India*. New Delhi: Indian National Science Academy and Indira Gandhi Rashtriya Manav Sangrahalaya.

Mehta, Lyla, Melissa Leach, and Ian Scoones (2001). "Environmental Governance in an Uncertain World", *IDS Bulletin* 32(4): 1-9.

Ostrom, Elinor (1990). *Governing the Commons: The Evolution of Institutions for Collective Action*. Cambridge: Cambridge University Press.

Ramachandran, K.K. and C.N. Mohanan (1991). *Studies on the Sacred Groves of Kerala*. Thiruvananthapuram: Centre for Earth Science Studies.

Steins, A. Nathalie, Niels G. Rolling and Victoria M. Edwards (2000). "Re-'designing' the Principles: An Interactive Perspective to CPR Theory", Proceedings of eighth conference of the *International Association for the Study of Common Property*, Bloomington, Indiana, USA, May 31-June 4.

Steins, A. Nathalie (2001). "New Directions in Natural Resource Management", *IDS Bulletin* 32(4): 18-25.

Wade, Robert (1988). *Village Republics: Economic Conditions for Collective Action in South India*. Oakland: ICS press.

8

Evolution from Thengapalli[1] to Community-based Ecotourism

Dhani South Panch Mouja Jungle Surakshya Samiti

PANKAJ SHARMA

Introduction

It is a recent phenomenon that consequences of abrupt climate changes questioned the very existence of mankind, resulting in attention of policy-makers towards management of common property resources (CPRs) like forests, water etc. historical evidences wherein the government took control of CPRs gave dual results. It is both due to lack of financial and human resources required to manage vast resources and rapid exploitation of resources by the local communities, who lost ownership over them. Orissa is one of the few states in India which still inhabit biologically diverse forest ecosystems. Orissa is also one of the states where community conservation efforts have given fruitful results. According to one estimate published in Biodiversity Conservation Prioritisation Project (BCPP) there are at least 4,000 villages in Orissa (total number of villages in Orissa 60,000) which are protecting over 2,50,000 hectares of forest (approximately 5 per cent of the total forest area of Orissa) (Singhs, 1993). But there are only few community managed initiatives which can be cited as successful examples, Dhani South Panch Mouja Jungle Surakshya Samiti (literally: Forest Protection Committee) being one of them. Dhani Reserve Forest which consist of 6 compartments, out of which compartment V and VI are protected by the Dhani South Panchmouja Jungle Surakshya Samiti was formally constituted on 1.10.1987 by 5 villages known as *'Panch Mouja'*. Dhani is synonym for community protected forest which is spread over 839.75 hectares.

---

1. *Thengapalli* is a system of local governance which involved physical transfer of wooden stick among different households which is also a transfer of authority and responsibility. This system was adopted by many villages in Nayagarh district of Orissa during forest protection initiatives where wooden sticks were kept outside the house of people who are supposed to patrol the forest for the following days on rotation basis.

The five villages which constitute *'Panch Mouja'* are located in Ranpur block of Nayagarh district in Orissa which is near to the northern part of Chilika lagoon. It is situated in 85°19'30' and 19°56' N at a distance of 73 km from the capital city Bhubaneswar, 12 km away from Tangi (district Khurda) and about 80 km from district headquarter Nayagarh. Ranpur was one of the princely states that merged in Orissa, after Independence on 1st January 1948 and formed a part of the district of Puri.

IUCN (International Union for Conservation of Nature publication *Communities and Forest Management* included case study of Dhani as the only example from Asian Continent. Dr. Madelene Ostwald at Centre for Climate Science and Policy Research, Linkoping University (Sweden) studied Dhani forest as a part of her doctoral thesis.

Genesis of Dhani South Panch Mouja Jungle Surakshya Samiti

Initially Forest Department along with Orissa Forest Development Corporation (OFDC) was involved in leasing out the forest to contractors, as a consequence, villagers felt insecure of loosing Dhani forest resources. They also started exploiting timber at a greater pace leading to depletion of forest and wildlife. Similarly increasing population, forest clearing for shifting cultivation commonly known as *podu* were few other causes which had their adverse effects on life of the villagers inhabiting *Panch Mouja*. Harmful effects of forest depletion like decrease in water table, soil fertility, rainfall, fuel wood, timber etc., made few villagers from Barapalli and Arjunpur to mobilise community and constitute the committee.

Table 8.1

Timeline of Forest Protection

Period	*Activity*
1961	Abundance of forest.
1961-62	Coupe given to private contractors by Forest Department (FD) . Coupe given to Orissa Forest Development Corporation (OFDC).
1972	Scarcity of forest produce.
1982-85	Combined meetings of five villages facilitated by notable individuals of Arjunpur and Barapalli.
1987-88	Forest protection committee was formed.
1989	Watcher was appointed. The committee got registered.
1992	The committee took the lease of cashew plantation to raise forest fund.
1993	Paid watchers appointed.
1994	VSS formed by the FD which became the first ever VSS in the whole of Tangi range.
1995	Potentiality survey of bamboo by FD. Ecorestoration scheme carried out by Social Forestry Project.
1996	VSS got approval from FD and management plan was done. Rs. 33,000/- was allotted for forest management work as a part of the microplan.

Note: VSS-Van Sanrakshan Samiti.

Source: Biodiversity Conservation Prioritisation Project. Vasundhra, Orissa.

Evolution

Forest protection committee inculcated improved forest protection policies and management practices. They tried to make committee decisions more democratic and spread the message of conservation to the next generation.

- Decentralisation of powers: initially only the elderly people of the village were given the rights to be the members of the committee but subsequently the membership opened for all.
- Environmental awareness: committee organises environment awareness in the village school sensitising children towards forest resources. Occasions like Wildlife Week, Environment Day are celebrated in village school premises.
- Legal body: After 10 years in 1997, Forest Department supported the committee which gave it a legal status and was registered as Van Sanrakshan Samiti (Forest Protection Committee) under joint forest management (JFM).

Local Governance

Core committee consists of 25 members selected by the villagers. They have tenure of 2 years. The core committee has representation from each village and from both genders. Both forest ranger and divisional forest officer (DFO) are notified about the newly selected committee. The committee maintains its own 'cashbook' and 'resolution book'. It selects its office bearers as 'president, secretary and cashier' for smooth functioning of the committee. The members once selected can be reselected for next term based on their performance.

The committee members meet on monthly basis. But in case of need, emergency meeting can be called. Executive committee takes decisions but if they are not able to take decisions general body meeting is called wherein representative from each household participates in the decision-making.

Modus Operandi

In the beginning *Thengapalli* system was followed wherein 10 members from each village used to go to the forest for patrolling and next day morning keep the *thenga* (wooden stick) in front of next person's house. Wooden stick reminds the *pallia* (person on duty for patrolling) for his turn to go to the forest. Later when they got enough funds they appointed two people as watchers (committee appointed forest guards). Committee members have the right to surprise visit to the forest for vigilance. Now under JFM, Forest Department provided one regular forest guard.

Managing through Sustainable Practices

In the early years of forest regeneration all activities like grazing of domestic animals, collection of timber and fuel wood collection was restricted which was given relaxation after some years. Now Villagers need to apply to committee for utilising forest resources clearly indicating purpose and amount. Applications are sent to the committee which are discussed in the committee meetings and approved if found genuine. Approved resources are to be utilised only for the household work and can not be sold in the market. Head-load of dry trees and twigs can be collected as fuel wood. Use of cycle or other vehicle for carrying fuel wood is not allowed. Hunting wildlife is strictly prohibited.

Awards and Recognition

Dhani is known for their commitment to forest and wildlife which gave them much recognition in India and abroad. Few of them are listed as below:

- *Prakriti Mitra* (nature lovers) award in 1998 (trophy and Rs. 5,000/-).
- Indira Priyadarshini award in 2002 from the chief minister of Orissa Naveen Pattnaik (trophy and Rs. 1,00,000/-).
- Committee received Rs. 1,00,000 as cash prize and honoured as Best Van Sanrakshan Samiti (VSS) 2009 in whole of the state.

WWF (World Wide Fund) with the help of IISc (Indian Institute of Science, Bangalore) and Vasundhara—an Orissa based NGO studied biodiversity at Dhani under BCPP. A documentary was made on community efforts which are showcased in different training programmes designed for the forest officials.

Resources

Generating resources was one among the biggest challenges initially. Even paying for committee forest watchman was not easy. Villagers came together and decided to contribute one handful of rice from each household which was given to the forest watchman.

- Government contribution: Forest Department realised the untiring efforts of committee and started helping them in cash or kind to sustain the community efforts. Forest Department is integrating the committee with government initiated development activities like Bamboo Mission.
- Re-leasing of cashew plantation: in the foothills of Dhani, cashew cultivation is done. It is one of the important livelihood activities in *Panch Mouja*. Every year it is given on lease to local contractor by the committee. March to May is harvesting season for cashews.

- Penalty: committee penalises individuals engaged in exploiting forest and wildlife. Accused is dealt as per the law of the committee and even handed over to forest department for serious crimes like wildlife poaching. Penalty may be reduced in case accused person surrenders and accepts the guilt. In a recent case they collected penalty as high as Rs. 35,000/-.

Challenges

Committee has to face a lot of challenges to accomplish different activities.

- Future leadership: present youth of village who look for quick money do not find a sustainable future in managing forest reserve and choose more lucrative option like migration to nearby cities for suitable jobs.
- Resource generation: committee does not have a right to sell forest produce and generate revenue for the committee. Moreover employed forest guards since long are paid very less salaries.
- Managing five diverse villages: five villages are heterogeneous in caste and culture. Committee has representation from general caste, scheduled caste (SC) and scheduled tribe (ST) population. SC and ST feel they do not have a say in the committee decisions and many a times boycott committee meetings.
- Mismatch between Forest Department and committee policies: there are instances where the committee feels a decision is good for the committee but Forest Department policies do not permit it. Recently there were gregarious bamboo flowering in Dhani forest which is a rare phenomenon and require bamboo cutting otherwise there may be a threat of forest fire but even after eight months it was not done.

Community-based Ecotourism: Integrating Innovations with Community Efforts

Council of Professional Social Workers (CPSW) and Wild Orissa—Orissa based NGOs are implementing an ecotourism project to provide livelihood opportunities for villagers. The project is initiated with a notion that committee itself can manage the enterprise after two years. The committee selects beneficiaries who are given training in different areas like hospitality ecoguiding, bamboo craft etc. Biodiversity, community efforts and local culture are being documented which can act as tourist attraction. The project is also supporting marketing efforts to attract tourists and at the same time framing policies so that the place can follow principles of ecotourism i.e., not hampering the fragile ecosystem and helping the local community. Revenue generated through the ecotourism will be owned by the committee which will be utilised for meeting operational expenditure and maintenance of forest ecosystem.

Innovative Approach

- Ecosystem approach to livelihoods: conservation of forest and biological biodiversity are one of the objectives along with generation of livelihood for the community. It is an effort to apply ecosystem approach and linking long efforts of community with livelihood.
- Community-specific solution to livelihoods: as each community is a unique case in itself, so should be the solution. Community based ecotourism at Dhani is also an effort to provide solutions specific to community and not stereotyping them to few solutions. Later other supporting interventions like training in backyard poultry, bamboo craft making and dairy farming are included for increasing beneficiaries and linking it to project and local markets.
- Community owned venture: the project aims to support the community in capacity building so that people themselves can manage the venture after two years. Both technical and soft skills are imparted to community selected beneficiaries in purview of the project. Technical support includes building up infrastructure with the help of transfer of technology to local community wherein community is involved in making bamboo huts, watchtowers, training in managing different records, knowledge of wildlife etc. Soft skills involve training in hospitality and communication to provide delightful experience for visitors.
- Multi-stakeholder approach: CPSW is implementing the project where they have to do capacity building and marketing support. Wild Orissa (NGO) is helping in wildlife documentation and developing information and communication material. Forest and Environment Department, Government of Orissa is acting as policy regulator and technical advisor. Current approach has created a platform for knowledge sharing and better understanding of policies which needs to be taken care of during the implementation.

Conclusion

Managing CPRs need a local institutional framework supported by corresponding government department and not vice-versa. Devolution of powers from forest department to formal community based institutions not only helps in conservation but also attract other players like NGOs, corporate to help community in including innovative practices like community based ecotourism, organic farming etc. There are few examples of community initiated sustainable practices to manage CPRs which need to be supported to rejuvenate the interests of new generation. Initially local governance needs to be strengthened to help people in setting up institutions and making people aware of various policy regulations through constant support and feedback, which is also helpful in decision-making. Policy dialogue while formulating forest policies should involve communities.

Dhani South Panch Mouja testifies that CPRs offer both opportunity for collective access to consumption and collective responsibility for maintenance of these resources. Active involvement of local communities in use and protection of CPRs can help in sustainable forest management. The sustainable use of these resources can generate adequate livelihood opportunities for the community.

References

Poffenberger, M. (1996). *A Report of the IUCN Working Group on Community Involvement in Forest Management.* Cambridge, UK: IUCN Publications Services Unit.

Ukil, S.C., G. Rao, R. Panigrahi and A.K. Nayak (1998). *A Decade's Experience of Dhani Panch Mouja Jungle Surakshya Samiti: Biodiversity Conservation Prioritisation Project.* Vasundhra Orissa. *http://www.vasundharaorissa.org/Research%20Reports/Conserving%20Biodiversity%20A%20Decade%27s%20Experience%20of%20Dhani%20Panch%20Mouja%20People.pdf*

This paper is based on working with five villages which constitute Dhani South Panch Mouja Jungle Surakshya Samiti. I would like to specially mention Mr. Dand Pani Das (Former Committee President), Mr. Benamara Das (one of the founder of Forest Protection Committee) for their continuous support during the work.

III

Governance, Laws and Industry

9

E-governance of the Forest

Management by Multivalence

DAVID MICHAEL COOK

Introduction

Poor forestry outcomes are continuing to occur as global, national and local authorities struggle with key governance issues through poorly enforced regulation, asymmetrical authority and control, and embedded corruption. However the last decade has seen widespread reforms and shifts in authority. These shifts are characterised by the move in authority away from government towards governance. This new form of governance, often termed 'new modes of governance' represents a shift towards greater participation, information sharing, and stakeholder currency amongst all actors. Within this context e-governance emerges as an obvious vehicle towards the achievement of improved forestry management outcomes.

Forestry in the developing world has embraced new modes of governance in the pursuit of forestry protection, reduced emissions from deforestation and degradation (REDD), and the reduction of illegal forestry. The role of regulatory governance takes on new meaning by positioning governance as an overarching solution that incorporates not only a good government, but also key participation and interaction between remote and rural stakeholders, public and private sectors, and their various networks, national cooperations and global interactions. These interactions involve not just formal pathways but also informal arrangements. They also signal the advent of participation by private actors and not simply public sector management. The swing towards including a range of information and communications technologies (ICT) and e-governance mechanisms signals an enormous advance in the development of governance in the global south.

At the same time one must understand the enormous shift in stakeholder arrangements in forested areas. Forest-dependent poor previously existed as stakeholders suffering the impacts of greedy government cronies such as

Indonesia under President Suharto (McCarthy, 2007: 3), and Pakistan with the princely state of Swat (Kruseman and Pellegrini, 2008: 10). Forest-dependent people are now more active stakeholders/participants within the better governance of a decentralised and more democratic forestry debate. However it is the nature of that participation and the interaction with new private actors and stakeholders that is pursued in this discussion. More specifically this hypothesis looks at the increased involvement of private actors in e-governance by examining emerging communication upgrades and the rollout of real-time spatial mapping in the regulation and authority of local forestry issues. The developing world has its fair share of both corporate and individual miscreant private actors.

Overall, those nations in the global south with forested areas suffer more than their northern counterparts. Indonesia has for decades anticipated the promise of basic governance reform in the form of regional autonomy. In 1957 Sukarno's centralised government reacted to very strong regional opinion and public outcry by enacting new legislation on regional government (Law 1/1957) that pointed towards decentralisation and regional autonomy (Legge, 1961: 168). However, less than two years later the President (Sukarno) brought about his version of 'guided democracy' and the notion of regional and local governance suffered a significant setback (Eckardt and Shah, 2006: 234). By 1967 the 'new order' regime of General Suharto further centralised the political system of Indonesia and led Indonesia into further 32 years of authoritarian rule. (Ricklefs, 2001: 356) The regional autonomy law of 1974 (Law 5/1974) did nothing to reverse the wait for regional governance (Jaya and Dick, 2001: 222) and it was not until Law 22/1999 under President Habibie that any genuine expectation of governance reform in remote and rural areas seemed possible (Eckardt and Shah, 2006: 235). In September 2004 the Indonesian parliament endorsed Law 32/2004 on subnational governance, and this legislation has cemented a genuine and purposeful intention by the greater Indonesian political authorities to embrace workable regional governance.[1]

Regulatory governance craves greater levels of participation than traditional government hierarchies (Peters, 1996: 8). When applied to forestry in remote locations and amongst stakeholders of proportionately asymmetric differences, this governance reveals transparency problems, limitations in communications, and difficulties in cultural understanding between all parties. Decentralisation creates the need for complex stakeholder involvement in remote and rural

1. Law 32/2004 draws some criticism from scholars suggesting it rolled back decentralisation by withdrawing some of the powers of the Regional People's Representative Council (DPRD) members (McCarthy, 2007: 11). However more importantly it allowed for the democratic elections of District Head (*Bupati*) and the Governor (*Gubernur*) (Diprose, 2007: 1) and enabled autonomy to be better understood (Wollenberg, Moeliono and Limberg, 2009: 18).

areas where previously local and indigeneous stakeholders were ignored. The notion that Indonesia's democracy is now established, transparent and stable still ignores the plight of indigeneous forest-dependent stakeholders. Despite the speed with which Indonesia's democratic processes develop, not a single political party has developed any clear environmental policy (*Jakarta Post*, 2008). In the 2001 election the leading five political parties had no environmental policies whatsoever (SBS, 2002: 356). This is an important factor when analysing modes of governance because in Indonesia environmental policy development, and accompanying environmental governance, is advancing independently from party politics. Instead it develops largely through connections at the international stage or through local pockets of interest. Indonesian NGOs draw support from other NGOs around the globe. Their main tool in the promotion of environmental governance in Indonesia is not through more traditionally western lobbying methods, but is more often through the dissemination and promotion of environmental awareness through ICT (Suan, 2003: 376; Hill and Sen, 2005: 39).

The role of Indonesian e-governance takes on greater significance in forestry for two interesting reasons. Firstly, forestry issues in Indonesia are characterised by a lack of party political and parliamentary debate and discussion, leading to the continued acceptance of flawed logging legislation and continued blurring between legitimate logging operations, forest resource management, and illegal logging practices. Secondly, the increase in involvement of forest stakeholders in the governance of Indonesian forestry relies more and more upon electronic communication mediums for its participation, authority and endorsement (Gregersen and Contreras-Hermosilla, 2001:240). This involvement goes hand in glove with a new generation of private actors in environmental issues such as forestry. The emergence of carbon credits as a rising form of international currency, coupled with the transfer of global attention from logging to carbon sequestration and the growing awareness of the economic advantages of carbon credit schemes, have moved the focus of attention onto AD (avoided deforestation) and REDD.

E-governance implies at least one of two levels of technology. In more general terms it usually refers to widespread computer and internet usage for the purpose of facilitating transparency, communication and participation to support governance. In a higher technology sense it refers to complex ICT technical solutions, such as satellite imagery, that permit the detection of illegal forest usage. When both these levels of technology occur concurrently within the participation and influence of a common group of private actors, e-governance takes on a far more significant role. This paper examines how the

Landgate Firewatch[2] project supports the theory that e-governance can be a genuine mode of new governance, and how private actors can and do influence authority, power and participation despite their size, scale, or location.

It is important to consider governance not as an arm of government, but rather the other way around. Genuine multivalence requires "new thinking" about governance that has ceased to focus on hierarchical frameworks, bureaucracy and administrative processes. This idea requires multiple applications, multiple interpretations, and the acceptance of multiple values. In this format the focal point of governance is more concerned with the characteristics of decentralisation, steering and multiscalar participation. It also necessitates the acceptance of a legitimate role by private actors in governance, and their interaction with other stakeholders. This includes the way in which private actors are able to tease out soft law arrangements that are much more closely connected with small scale stakeholders, such as the forest-dependent poor. In addition one must accept that governments (or their departments and subdivisions) can and do act as private actors rather than in traditional roles of government. This includes governance-based arrangements between regional, remote and rural areas of authority as well as interconnections with global economic and market-based interests. E-governance can act as an enabler for these new modes of governance, which in turn creates many more opportunities for the introduction of good governance.

In this sense e-governance is indeed a legitimate mode of governance. It posits that ICT and high level technologies, when shared rather than controlled, can bring about successful and sustainable pieces of authority that bypass the hierarchical machinery of government, whilst engaging a wide range of stakeholders from vastly asymmetrical backgrounds. The specific use of spatial satellite data combined with local forestry knowledge to create highly compelling packets of information that can mobilise stakeholders is of great interest. It can allow for managed forestry outcomes that are beneficial to an increased number of stakeholders, yet do so without the heavy-handed enforcement of centralised authority. Whilst this is not to advocate that e-governance bypasses every level of Central government, it does suggest that e-governance can indeed germinate the conditions under which local and regional soft law arrangements have replaced many elements of Central authority.

---

2. The Landgate Firewatch Program is a project using Nasa space satellites, Australian spatial data software, and a network of Indonesian contacts involving a variety of electronic media including, satellite data, highspeed internet delivery, mobile phone communications, SMS messaging, and radio broadcasts. It is examined in this paper as an example of how e-governance can deploy complementary mechanisms to achieve high level governance outputs, transparency and participation.

New Thoughts on Governance

Governance refers to more than just a set of ruling guidelines. It is concerned with the systems, institutions and norms by which governments, businesses, associations and society are directed and controlled. Revised thinking on regulation and power proposes that it is crucial to consider the term 'governance' as a more powerful term than 'government' because it demonstrates an understanding of the ways in which influence, control and authority can protect order even in the absence of state activity. In place of state authority, governance seeks to emphasise the augmented role of non-state and private actors as well as facilitating a shift towards a variety of systems for managing networks, partnerships and connections.

Forestry governance depends on a multifaceted set of interactions. The moves to decentralise and regionalise formal and hierarchical government authority also help to fertilise local and regional demands. Globalisation, information technology, and increased communication speeds have also facilitated a much more powerful expanding role by international markets and NGOs. In Indonesian forestry for example, there are now many enthusiastic stakeholders. The forestry industry's governance is now driven by a range of factors including environmental, political, human rights, social, cultural and economic issues. This new governance takes its shape from a broad raft of hybrid and negotiated relationships that intersect between private and public sources. Consequently this form of governance is less about state-centric authority and more about multiple stakeholder interactions within the global marketplace. It attempts to protect and regulate public and private interests that range from local and regional scenarios upto global and multinational settings.

A Different Perspective on Governance

New modes of governance require an adjustment of thinking about what we term governance. This begins by looking at government as only one wing of governance, rather than the other way around. It sees governance as a multilevel authority that has a much more complex set of interactions than a central 'single point-form of government' (Walters, 2004: 34) and it looks at the multiscalar characteristics of self-regulating cross-boundary networks, 'steering' (Osborne and Gaebler, 1992), and the effect of 'hollowing the state' (Rhodes, 1996).

Traditional understanding about governance was previously centred on hierarchical frameworks, bureaucracy and administrative processes, and their timely applications. As globalisation became more prominent in transnational thinking, (Speth and Haas, 2006: 9) and as the business of government became more complex (Walters, 2004: 30), governance shifted its emphasis in line with neo-liberalism to align with the new requirements of deregulation and privatisation. The term 'governance' shifted in its meaning to incorporate transparency, accountability and cross-boundary networks of varying size and scale, capable of organising themselves rather than being completely dependent upon the management processes of state-centric top-down hierarchies. Whilst the broad task of managing forests still has many links to formal government institutions, decentralised intersectoral networks are becoming more evident. New governance literature (Sassen, 2003a: 2; Walters, 2004) notes that there is a much more complex version of governance than either market-driven transparency or government hierarchy.

There is a much broader application for governance that is multiscalar insomuch as it places a much greater emphasis on a wide range of magnitudes amongst stakeholders (Sassen, 2003a: 3). The notable shift in governance in this sense is its application not only in economically powerful transnational interactions, but also in a broad array of local, regional and subnational settings that occur at the same time as larger global marketisation. Governance now requires an understanding that focuses on more than just matters of global scale. It now deserves attention across a range of accompanying local and subnational connections (Ibid: 3). For forestry governance and property management to have genuine effect and meaning it has to extend its parameters to include interactions of varying scale and allow for different notions of authority, participation, values and expectations.

This idea of varying scales is particularly synergetic with an understanding of digital technologies. Electronic markets, internet knowledge archives, social networking, SMS messaging and a range of utility logic systems can all represent interconnecting actors from both private and public spheres (Latham and Sassen, 2005). They allow individuals, corporations, NGOs and governments to converse, share and distribute both information (data) and social intention. They permit governance to permeate from one actor to another, irrespective of the actor's size, location and authority. E-governance might stand for 'electronic' governance in one sense, but it also most definitely stands for 'enabling' in another (Cheong and Yong, 2003: 394).

There are various luminaries who contest the traditional notions about governance beyond Sassen's ideas. Rhodes (1996: 653) puts forward not one but six differing conceptions for what might be used to describe governance. He identifies that rather than looking at these characterisations as examples of ambiguity, they should in fact be seen as indications of the heightened desire to promote the role of governance to a more universally applicable state. Governance in this sense has emerged through changes in globalisation, technological advancements, and exponential improvement in communications.

These changes have allowed individuals, groups and corporations to cut across systems, borders, hierarchies and nations whilst assumedly more powerful organisations such as governments have failed. This is a fundamental characterisation of the way in which common property in the form of forestry has been mismanaged.

Two of Rhodes' main conceptions, 'socio-cybernetic systems' and 'self-organising networks', are therefore of particular interest because they resonate very closely with the notion supported by this proposal; that digital communications and ICT are of immense value in developing good governance. Both of these descriptions fit comfortably with the ubiquitous nature of ICT and the way in which information and discussions are shared freely and rapidly with stakeholders without the need to pass through a governmental hierarchy or system of authority.

In this new broader understanding of global and local interactions then, it is governance that assumes a much higher mantle than government. Contrary to early 20th century 'old governance' thinking (Walters, 2004: 29), government should now be viewed as simply one arm of governance rather than the alternate. The advancements and improvements that accompany globalisation are also partnered by a similarly vast increase in complexity of government, regulations, international differences, and local, regional and sub-national distinctions. These multiscalar distinctions also carry with them the need to understand social and cultural complexities. The Indonesian forest dwellers of the early 20th century were considered by their Dutch colonial overlords as nothing more than slaves (Steinberg, 1987:158). They had no stake or say in the allocation and harvesting of forestry resources. Similarly during the New Order the 'forest dependant poor' had no voice in the systematic destruction of forests by illegal logging companies. Today the forest dweller and his tribe/village/family/community are important stakeholders in a highly complex set of engagements that are globally networked and characterised by multiple organisations all seeking to self-regulate. In this sense it is easy to understand why Rhodes posits that we might consider the notion of 'governing without government' (Rhodes, 1996: 667).

The terms 'scale' and 'multiscalar' therefore can refer to the size, dimension and authority of interconnecting actors, yet can also refer to the size and scale of the complexities of the issues, the cultural disparities of each issue, and the 'dynamics through which relations and domains are constituted' (Latham and Sassen, 2005). Pockets of authority that previously 'nested' within local scales now move freely into more global trans-boundary scales, whilst globally positioned power bases now venture into local areas which are culturally and asymmetrically complex (Sassen, 2003b: 2).

Rhodes sees governance as 'self-organising, inter-organisational networks' (1996: 652) that are complementary to the pathways that are sought in the allocation and distribution of resources. Governance provides authority to manage networks and interactions in a way that is free from the Central authority, embraces participatory management, and carries legitimacy that is acknowledged both locally and externally. This new system of authority is not simply a decentralised government. It is a highly interdependent sociopolitical arrangement that incorporates NGOs, local and regional voices, corporations and public and private actors. This form of governance derives its legitimacy from dynamic and interactive set of exchanges among all the relevant actors. For Rhodes, the more blurred the boundaries, the more shared the goals and the more interdependent the administrative connections, the more legitimate the governance.

This renewed version of governance provides a range of pathways in which social, political and economic life can be administered without a heavy dependence on government bureaucracy. Rather than bemoan the complexities of global interaction, and by default the complexities of government, new governance thrives on complex interfaces that provide multifaceted opportunities for engagement. Complex issues give rise to pluralism across different levels, locations and structures. In particular e-governance maintains and develops these new pathways. Many observers suggest that ICT is such a profound force that it has 'outrun and bypassed the machinery of government' (von Furstenberg and Kirton, 2001: 244). Rhodes' inclusion of sociocybernetics as a key dimension of e-governance is important because in large and asymmetrical social systems most people never see each other face to face but instead communicate using a variety of other means such as email, internet, SMS messaging, social e-networks and telephone. 'The issue of whether someone lives in a centre or on the periphery of a social system is determined less by his or her place of residence but rather by the intensity and the connections of his or her communications' (Paetau, 2003: 30).

Governments have had to pass multiple, complex laws and regulations to enable foreign firms to operate in different countries. This process has contributed to the destabilising of established state power and authority (Sassen, 2003a: 8). In its place has emerged a significant increase in various forms of private authority that have forged their own set of institutionalised and interconnected relationships. There is a difference in the size and scale of each actor, so that even though a network may be global in its connections, it does not have to exercise its governance on a global scale. The advantage of multiscalar connections is that representations can occur locally, yet do not necessarily rely on a slow progression upwards through 'nested' scales to pursue a particular struggle. Instead, any actor can jump from local to international involvement regardless of size, country, border or jurisdiction.

By the same connections that allow small local groups to interact with large private firms, transnational companies (TNCs) can put large resources into making whatever formal or informal networked connections they deem useful. TNCs work around or over governments in whatever directions combine best between their profits and their compliance to regulations, authorities, interests and concerns. They do not fear complex arrangements, but instead look to forge interconnections with any actors, regardless of scale, that seek to engage them within the scope of their activities. In a less formal way than through the channels of highly regulated government, the actors in these self-regulating networks are subject to governance that is fluid and dynamic in nature. It is a shift from 'institutions' to the 'processes' of rule (Walters, 2004: 29).

Steering

An additional shared characteristic of new governance is the depiction of 'steering' rather than 'rowing'. Steering has significant undertones for governance because it can imply a certain technical determinism that could take away political competition. Steering might instead suggest that there is only one direction and that all parties should follow that direction. This notion is more concerned with directional guidance rather than being a force to propel actors to behave in a certain way. Global moves to privatise banking, and the subsequent 're-regulation' by controlling governments is one such example. This form of steering allows for a much greater range of dynamics and participation. This has removed much of the politically-determined nature of finance and lending issues,[3] and replaced them with more regulatory focused meetings that allow for a steering process to evolve in areas such as lending for microenterprises.[4] The benefits in terms of forest-based small scale activities are significant because each enterprise invigorates much greater levels of participation from low-income forest-dependent people, which, when coupled with low-cost e-governance solutions provides for significantly richer engagement, better outcomes, and far more resourcefully widespread forest management (FAO, 2005).

---

3. The Indonesian Banking Architecture Programme (API), banking-system was agreed to in 2003 and launched in January 2004, although much of its rollout is a result of World Bank and International Monetary Fund (IMF) undertakings rather than an independent attempt to introduce world-class financial regulation. Nevertheless it has been of enormous assistance in the stability and growth of small and medium enterprise lending such as is required in remote and rural parts of Indonesia. It has contributed in facilitating the less formal governance that drives many of the successful microlending schemes that are recent developments in Indonesia over the last eight years.

4. It has also sought to prevent any further recurrences of the corrupt banking practices that drove forest sector mismanagement in the 1990s and in particular the selling of forestry debts during and after the 1997 financial crisis (Barr, 2001).

New forms of governance offer a type of steering that is highly participatory, where those doing the steering are networked to each other in a constant state of renewal. The process is dynamic, participatory and carries all of the socially embedded characteristics of the private sector, without losing its connection with the public sector. It involves the persistent re-examination of government and intergovernmental institutionalism, whilst aligning authority according to norms and values. This process tends to be informal, highly social and highly interactive. E-governance, through social media such as email, SMS, and website interaction, is a crucial enabler in allowing steering to determine its directional pathways.

The problem with accepting notions of steering in new governance is that it affects the ways in which government authority is viewed. The term 'hollowing of the state' (Jessop, 1997: 578; Rhodes, 1996: 661) refers to the state in decline. It suggests 'zero-sum' thinking whereby the increase in authority through informal new forms of governance must imply a corresponding decrease in authority by the state. Governments that fear their authority is declining have reacted by shifting some of their authority towards meta-governance. This is a higher level of authority that allows seemingly displaced government institutions the opportunity to retain some hierarchical control by restructuring themselves 'over' emerging new forms of governance rather than trying to compete with them.

Whilst the steering effect that has evolved in new modes of governance is greeted with enthusiasm, the steering by governments in the form of meta-governance is viewed with caution. The 'shadow of hierarchy' is a legitimate criticism of the struggle of those seeking to secure hegemony (Jessop, 1997: 575). However, the simple truth is that despite the appeal of new modes of governance there are significant numbers of multinationals who turn to capital flight, relocation and extra-territoriality to avoid the constraints of authority and regulation. Governments move towards meta-governance in order to retain some control over lost governance from new modes, and also deploy their own 'shadows' as a means of reintroducing some political influence into privately self-regulated activities that may have bypassed a political agenda in what has been coined as the 'antipolitics of governance' (Walters, 2004: 33).

Forestry management suffers from a significant disparity between its various actors. One of the genuine benefits of steering by new forms of governance is that private actors have the knowledge and expertise to steer in the most useful and beneficial directions because their comprehension of each issue is more fundamentally grounded. However, there are occasions where governments use meta-governance to retain control of the institutional direction of a particular area. Heritier and Eckert (2007: 4) note that in cases of environmental protection, there is a large enough asymmetry of knowledge and capital between the expert private actors and the state, that a range of politically compromised incentives and disincentives emerge as a means of procuring the state's preferred institutional outcome, rather than a set of networked interactions that evolve a more apolitical self-regulatory result. In the case of forestry management in developing nations there are at least three competing sets of actors and associated knowledge spheres. The timber trading industry, the government, and the forest dependent locals, all lay claim to expertise in the direction and guidance of forestry issues. The significant difference is that local indigeneous representatives lack both the power to pursue interests and to construct 'knowledge' accordingly, and at the same time lack the sophistication and the financial clout to properly compete on a knowledge basis with international timber businesses. Clearly if 'steering' is to navigate a steady course, the speed, clarity and omnipresence of an enabler such as e-governance is required.

Forestry meta-governance in this sense comes from groups such as the Forest Stewardship Council (FSC) through their certification of sustainable timber (Visseren-Hamakers and Glasbergen, 2007: 410). The FSC is essentially a global private intersectoral partnership that brings together industry, social groups and environmental groups.[5] A key element in this type of meta-governance is the attempt to share the balance of power between the three groups from social, economic and environmental sectors. However, the FSC's effectiveness depends upon both the public awareness of timber product consumers and the preparedness of governments to enact laws regarding certified forest product (CFP) trade[6] (UNECE, 2006: 98). If forest management is to depend upon meta-governance for the greater part of its course-plotting then it must also look to a system that smoothens the progress of information sharing and the adoption of compatible values and norms. Again, as with any workable form of governance, there is a strong need for shared, up-to-date exchanges between any and all stakeholders and forestry actors.

While most of the new governance literature has a sanguine theme, there are several questions that require consideration. Cutler *et al.* (1999) oppose the idea that governance emerges naturally from multiscalar interactions between private and public networks. They acknowledge that these interactions occur, but disagree with the manner in which their authority and legitimacy takes shape. Rhodes, Jessop, Walters and Sassen see state-centric government and governance as displaced by self-organising private interests. However Cutler *et al.* (1999: 4) maintain that for the most part the governance role has been

5. Note that the success of the FSC's program to distinguish between legally harvested timber and illegal logging is strongly contested.

6. Although the world's first forest certification took place in Indonesia in 1990, Indonesian consumers are not proactive in asking the source of the wood that they buy, or what impact that may have on indigenous forest communities. (Rainforest Alliance, 2007: 1).

voluntarily abandoned by the state. They advocate that governments (in some instances) have been unwilling to cooperate and have shown disinterest in extending their domestic regulatory capabilities outside of their own territories. This is of particular interest in determining new developments in regional and local authority with regard to forestry issues in remote areas of forest. States deliberately facilitate the involvement of privately organised networks so as to provide privately organised governance in lieu of their own. If this is the case, and private self-regulatory networks have inherited their governance role by default, it calls into question the notion of authority in terms of their legitimacy. Cutler *et al.* do not suggest an absence of either authority or legitimacy, but rather that different forms of privately networked authority have differing types and levels of legitimacy. A partial solution to these concerns comes from an ability to provide information and shared discussions between government institutions, regional authorities, local forest stakeholders and larger private corporate stakeholders. E-governance fulfills the machinery component of this solution.

Where modern governments appear unwilling to intervene in the authority of global commerce, and where international organisations appear ill-equipped to govern effectively, private firms are establishing international networks and associated groups that broker rules and standards for themselves and others. They are taking on the mantle of authority to govern in specific areas. Even when they are organising themselves to prevent governments from disrupting their business, they are acting with authority, legitimacy and power (albeit somewhat contested). They are doing internationally what nation-states do domestically, which is to govern. Whether the governance is enacted by a state government within a sovereign territory, or by an internationally multiscalar organisation across multiple boundaries and geographies, the governance is genuine if it is derived on the basis of trust, institutionalised standards, and cooperation. Good governance involves steering rather than rowing and requires multiple participants, stakeholders and a range of actors from different levels, scales and backgrounds.

'No single actor, public or private, has all the knowledge and information required to solve complex, dynamic and diversified problems; no actor has sufficient overview to make the application of needed instruments effective; no single actor has sufficient action potential to dominate unilaterally in a particular governing model' (Kooiman, 1993: 4).

New governance refers to an institutional shift that floats across all levels of government and through hierarchies at local and global levels. It sees government take less of a hands-on role and more of a guiding function in order to steer and regulate markets, private actors, NGOs and civil society (Bevir, 2009: 29). The distinction in authority and power between the state and civil society is blurred. New networks emerge, new interactions develop, and private/public partnerships take shape between actors in a variety of levels and scales. Private actors, NGOs and civil society now interact to form new networks, arrangements and partnerships that create their own authorities, and turn to markets and their social constituents for legitimacy and sustainability in the distribution of authority, rule and power.

Private Actors in Governance

Based on the ideas of this new generation of governance as explained above, democratic authority, private actors, economic development, and forest management share a crucial affiliation. Economists, and not just neo-liberals, believed that for market reforms to be effective there was a requirement for some oversight by political institutions. For governance to be of good quality it needed a level of economic reliability that placed an emphasis on efficiency, transparency and democracy. The involvement of private actors in new modes of governance therefore underpins the important connections to market reforms and market efficiencies. The term 'good governance' came about largely as a descriptor for the requirements of the World Bank and the IMF where the deployment of loans and aid money into developing countries needed some form of measuring stick for the legitimacy and authority of governments. Typically it was used to delineate between honest government and corruption, however it developed into a much wider definition that quickly sought the confidence of those outside the government, notably markets and NGOs. Good governance looked for distinctions between those who held power in the judiciary from those in legislature, and those in the executive.

The *Trias Politica*[7] became the first criteria for considering whether a country had good governance (Heywood, 2002: 315). However new governance has developed beyond Montesquieu's original ideas when describing strong governance and strong civil society. For agencies such as the World Bank the connection with good governance was made through linking strongly to new public management (NPM),[8] and the process of privatisation of a variety of public sector operations in the pursuit of competition. Other international organisations such as the United Nations were more concerned with social

7. The *Trias Politica* refers to Montesquieu's tripartite system known as the separation of powers. It refers to a model of governance that separates the executive from the legislature and the judiciary so that each area is independent from each other and free from undue influence from the other.

8. The NPM direction of governance is one of the definitions proposed by Rhodes. Whilst this is a useful definition of governance in one sense—it fails in two key areas. Firstly whilst it seeks to engage in private sector principles in relation to competition, it does not sustain these principles over time because it does not maintain a private sector environment. Secondly, it fails to keep up with the sharper knowledge that private sector actors constantly update as part of their 'competitive edge'. NPM is slowly shifting towards DEG (digital era governance) with a greater emphasis on holistic guidance and digital knowledge management.

justice, participation and environmental protection (Bevir, 2009: 27). The combination of these two types of governance highlights an intersection that cries out for an explanation of the role of private actors and the development of non-government based authority.

Private actors are therefore vitally important to a holistic model of governance. They restore the notion that governance is a multi-agency affair, networking with multiple actors in a shared project so as to engage in the business of governing (Clarke, 2009: 33). Under the banner of new modes of governance multiple agents are required because specific schemes can no longer belong to a single body (i.e., a government or government department), but should be shared between the competing and complimentary interests of a variety of different actors. Instead of private actor involvement implying that the state's authority is in decline, most researchers regard new governance as an effective partnership between private and public interests. The steering implied here is best described as 'co-steering' because it involves cooperation between many different forms of authority (Ibid: 34) from public sector at all levels through to private corporations, NGOs, minor stakeholders, and private individuals.

Examples of such governance in forestry are often described as public-private partnerships (Visseren-Hamakers and Glasbergen, 2007: 412) and many examples exist. The FSC and the Roundtable on Sustainable Palm Oil (RSPO) are both examples of private inter-sectoral partnerships that operate by providing meta-governance in one sense, certification at another level, and forest management between government departments (national, partnerships that operate by providing meta-governance in one sense, certification at another level, and forest management between government departments (national, regional and local), private corporations, education and training providers, individuals and civil society.[9]

Public-private partnerships (PPPs) have had a troubled past in the context of Indonesian forestry. The World Bank notes that poor financial regulation in Indonesia placed control of a number of monopolies and associations into a handful of New Order cronies, especially Bob Hasan (World Bank, 2006: 41). The Asian Financial Crisis of 1997 revealed the extensive banking mismanagement that saw privately-owned banks violate legal lending limits and government capital adequacy requirements in order to fund poorly regulated and selectively enforced forestry management. There are significant

9. The FSC and the RSPO are both public–private partnerships that run management frameworks in Indonesia that have some success but are both heavily criticised for their inability to curtail corruption and bad business practice. They appear outwardly to engage in all the right areas of governance—yet lack the real authority to enforce meaningful penalties to prevent illegal logging and land degradation.

public-private difficulties with governance in developing countries where corruption and collusion have undermined the management process (Barr, 2001). Many alliances continue to exist, trade and exacerbate the good governance efforts that are being attempted in forestry.

On the one hand researchers encourage the participation of private actors in forestry governance whilst on the other they are highly critical of negotiated soft law arrangements that are derived from varying levels of participation of multiple stakeholders both public and private. Undoubtedly private actors have a history of preying upon the asymmetries of local forestry stakeholders and poorly educated forest dependent groups. Three important points need discussing in relation to this. The first is that the asymmetries in the range of forestry stakeholders are so acute that the standard principles, norms and expectations of good governance are far less likely to get a firm foothold against the backdrop of well established financially powerful private actors. Under these conditions it is hard to foresee how to create a strong environment for sustainable local/regional forest management.

The second point is that there are a number of private actors from unlikely sources that change the forest governance landscape. These examples include where a government department or government sponsored agency appears as a private actor rather than as a branch of the public sector (UNEP, 1999: 1; Rametsteiner, 2009: 152). The Landgate Firewatch project acts and behaves like a private actor within the forest governance sector in Indonesia. The emergence of government departments in roles that appear to show them as private actors is now an established characteristic of globalisation.

The third point is that e-governance should be more closely considered because it is both an instrument of good governance, providing reliable and prompt information on forest management issues, and a vehicle for new modes of governance, creating a sustainable platform from which to anchor a range of informational, participative and knowledge-based assets. These assets provide powerful knowledge and forest control statistics that arrive in ubiquity. The value of e-governance in delivering broadly dispersed knowledge and information should not be underestimated. Information from web-disseminated satellite downloads cannot easily be filtered or blocked. Imagery of logging destruction is available directly from the source and instantly a part of the public sphere. If there is one area that looks capable of solving asymmetry, knowledge, and enforcement issues by enabling forestry governance, it is the inclusion of e-governance.

A key part in these 'governance-enabling' connections comes from the private actors. Private actors can create authority under both formal and informal conditions. In other words private actors might develop their authority independently from the state, or they might develop that authority through a closer relationship with the state (Cutler *et al.*, 1999: 335). Both paths can achieve governance that is derived legitimately. Similarly, that very same authority may emerge based on the displacement of state authority or it may emerge on the basis of market-based interconnections where no previous authority had existed.

Soft Law from E-governance

Soft law creates authority through informal conduits and less legally binding provisions. Soft law relies on private actors to drive these arrangements. New technologies and modernised approaches to networking, communications and interactions by private actors and the state come to the forefront of governance. The norms, standards and values by which we measure authority are now increasingly from a decentralised and informally constituted range of sources. Cutler *et al.* (1999: 346) note that some 'opponents to government regulation' have also controlled power and authority by way of the internet. These challengers to formal authority have a range of multileveled and multiscaled support from 'netizens'[10] that support new spaces for authority to emerge.

Knowledge and expertise can bestow a particular firm or private actor with authority. They can lay claim to authority on the basis of expertise, especially if it displaces a state or public authority that does not have the same level of proficiency. Once that knowledge becomes more distributed and resourced, it can be accessed at all levels. The more levels it devolves to, the stronger the private authority becomes. E-governance is a vital method in the universal distribution of knowledge, information, and hence authority. Knowledge and expertise enables the private actor's authority to emerge, and the regular application and distribution of the knowledge further empowers the authority, as well as institutionalising many of its characteristics (Cutler *et al.*, 1999: 347).

Private actors and firms are of interest in forestry since they tend to be more focused and more 'outcome-driven' than other organisations. As such they represent a more robust example of privately operated governance. They usually rely on ICTs for their development and global acceptance. The development of ISO standards globally provide many strong examples of how an authority operates in the area of industry norms, coordination services and private regimes. These groups typically integrate both public and private sector

10. A netizen (or cybercitizen) is a person actively involved in online communities. Netizens have a self-imposed responsibility to give and receive viewpoints, to furnish information, and encourage free speech and open access through electronic mediums such as the internet, SMS, email, Wikis and file sharing. They seek to do this across unstructured, hierarchy-free forums.

involvement (Cutler *et al.*, 1999: 356). As private authorities increase their institutionalisation at an internal level, they simultaneously broaden their wider authority and legitimacy. This broadening of their exposure and acceptance equates to increased levels of governance. Informal arrangements become ritualised over time, and then become institutionally confirmed. ISO standards are likened to a form of meta-governance, although they are far more closely related to hybrid regimes, because they evolve more fully from the interactions of private actors than the policy directions of centralised institutions. Nevertheless, they clearly demonstrate the creation of authority over a period of time.

Legitimate and sustainable private authority operates through obligation rather than through coercion. However the key issue for forestry governance is whether compliance occurs or whether the governance is ignored or bypassed. This is the true test of governance. Forestry groups such as the FSC may claim to exercise legitimate authority but the question needs to be asked: how legitimate is that authority to stop illegal logging when massive depletion is caused by large corporations? These private companies already have connections and relationships with private actors yet the illegal forestry continues. McCarthy and Warren (2009: 13) warn of the situation in community-based natural resource management (CBNRM) where some private actors become corporately aligned actors so as to secure their future over and above the future of their community. It is a difficult scenario. As local groups become more organised, they develop hierarchical structures (albeit at the local level) that create opportunities for some actors to cement a future for themselves and in so doing they become disassociated with the community-based cause that they mobilised for in the first place.

As private standards become adhered to, their compliance builds a form of authority. Such informal commitments are deemed to be interdependent. Within the same period of construction norms and values come into play, institutional legitimacy forms, and soft laws are discussed together with state authority. Private actors rely on regulation and enforcement to accompany participation, transparency and broad multiscalar networks. Such combinations work best when they take place across mediums that allow for a fair exchange between any and all stakeholders. E-governance is one such medium, allowing the same features of participation, transparency and multiscalar networks, whilst offering it in a highly accurate, expedient and trustworthy package.

However, private actors who enjoy authority may be swayed towards corporate actors over disadvantaged private actors such as forest dwellers. Soft laws and informally created authority can offer a less than equitable standard of governance to those with downwards socioeconomic asymmetry. Service providers that offer the 'promotion' of an industry standard, face a number of incentives and opportunities that skew favour (Freeman, 2000: 632). Privately organised forest management schemes are less likely to offer the same service and enforcement of agreed rules to forest dwellers as they are to highly organised multinational logging corporations. Concerns about imbalances in the ISO 14000 (management standard) planning also indicate that at the macro level of international governance, private actors are unsuitable soft law regulators in their own right, since they gladly accept state and publicly endorsed authority, but sidestep public and state scrutiny (Falkner, 2003: 75). Other researchers (Freeman, 2000: 640) maintain that this problem occurs with numerous NGOs and thousands of privately generated codes and standards.

The strongest private authority therefore seems to enjoy a hybrid connection between its public and private relations. The use of e-governance is typically the enabling factor in these hybrid arrangements. It can assist to exemplify the fluid state of governance between private and public actors (Cutler *et al.*, 1999: 369). The key distinction centres on the proportion of authority that is enforceable, accountable and still holds traction with both civil society and the marketplace (Ibid: 369). Cutler *et al.* suggest that private influence does indeed 'govern' many international elements through the legitimate exercise of governance. The more multileveled, blurred, cross-boundaried and multiscalar—the more legitimate the governance. Similarly for Freeman the real strength of decision-making relies on 'a variety of actors making collections of decisions in a web of relationships' (Freeman, 2000: 673). Thus e-governance potentially becomes more than simply a tool of good governance, but rather a vehicle for vastly improved forest management.

Market-Driven Characteristics

Despite the optimistic approach that frames these ideas about e-governance, private actors in hybrid connections need to do more than just turn up and participate. There is an important distinction between cooperation and governance. Governance may emerge from cooperation, but governance needs a long-term set of normative adjustments created and driven by private actors in order to come under consideration for the title of hybrid governance (Falkner, 2003: 73). It is important to acknowledge the pervasive nature of private actors and the ease with which modern technologies allow them to cooperate in multilevel networks. Private actors can be co-opted for a variety of market-serving reasons that might colour the legitimacy of their governance's involvement. Thus the blurriness that was for Rhodes a key ingredient in the legitimacy of private governance, is a problem of purity for Falkner. Amongst private organisations, NGOs are highly vulnerable, and their independence is under threat from the market.

The governance role that NGOs play in targeting private firms, particularly TNCs and large corporations, is problematic. Although NGOs have been highly successful in targeting specific companies, resulting in widespread environmental benefit (Falkner, 2003: 80; UNF, 2009: 45), there is a certain ad hoc quality that accompanies their successes. For every victory, there is a much greater number of failures where environmental damage, forest destruction, and illegal activities carry on without caution. NGOs suffer from the same survival instincts as the forest dependent poor that they might hope to protect. Accordingly, NGOs deliberately target areas of probable success and notoriety, rather than applying an even-handed approach to global environmental governance. Simultaneously NGOs increasingly getting reliant on corporate funding for their existence, may become domesticated towards their private and corporate sponsors. In the case of forest protection, however, there is a new source of funding that although corporately focused, offers private actors a range of sponsor and alignment strategies that run somewhat parallel to their own.

Indonesia's approach to new forms of e-governance and decentralisation shares the same complexity as many other developing nations that introduce governance reform. Management information systems (MIS) have been touted as the necessary technical upgrades to ensure good governance through e-governance. This is a mistaken approach, placing most of the emphasis on computer training, rather than training people to use technology to complement their participation and involvement. There is a distinction between management systems on the one hand and genuine interest-driven governance and involvement on the other (Kelly, 2006: 605; Nugraha and Rudiantoro, 2008: 92). At the smallest scale, where the forest dweller, villager, or peasant is initiated into new networks and associations, the issues of capacity building, education and empowerment emerge as requirements to ensure new governance. Projects that aim for good governance as a part of their framework contain gaps in methodologies and training materials that are essential for even small scale community-based forestry projects. Karousakis (2007: 35) suggests that there is a continuing didactic problem that requires a continuing commitment to training and capacity building courses. Corbera (2005: 48) goes further, suggesting that the marketplace does not permit ongoing education of the key grassroot participants because that would require significant effort on the part of ecosystem project managers and would introduce considerable ongoing transaction costs. However, the continuing trend towards the establishment of MIS for Indonesian NGOs takes the compensatory 'road' in terms of governance and fails to recognise the social interconnections that sustain e-governance involvement and commitment.

In simple terms NGO operations in local and remote areas for forestry are often given a computer and then left for an MIS-NGO to enter data and report operations. Compensatory measures such as this are weak examples of e-governance. They hold back the evolution of good governance through technology by avoiding existing social architecture and implanting a management system in its place. These MIS placements often simply collect and report information back to a Centralised organisation or NGO. MIS installations have, in many cases, simply switched Centralised government reporting to Centralised NGO reporting. This is not to suggest that compensatory and complementary forms of e-governance are mutually exclusive. Compensatory e-governance of this type is effective in authorising greater efficiency of a wide range of governance procedures. UNESCO[11] describes e-governance in terms of its administrative qualities by defining it as follows:

> Governance refers to the exercise of political, economic and administrative authority in the management of a country's affairs...[E-governance] may be understood as the performance of this governance via the electronic medium in order to facilitate as efficient, speedy and transparent process of disseminating information to the public and other agencies, and for performing government administrative activities. E-Governance is generally considered as a wider concept than e-government, since it can bring about a change in the way that citizens relate to governments and to each other. E-governance can bring forth new concepts of citizenship, both in terms of citizen needs and responsibilities. Its objective is to engage, enable and empower the citizen. (UNESCO, 2005)

In a holistic approach e-governance needs both compensatory and complementary forms of governance rather than 'mere technology exploitation' (Suan, 2003: 388). Since e-governance now tackles and solves many of the issues surrounding the complexity and size of administrative interactions, it is particularly important that those who support e-governance do not lose sight of its ability to imbricate in the social and human areas of small scale highly participatory local involvement.

Forestry stakeholders would do well to view e-governance as complementary, forging alternate pathways that can run parallel to existing

11. UNESCO is the United Nations Educational Scientific and Cultural Organisation. It allows for two-way communications between all stakeholders on a participatory 'pancake' that permits forest dwellers to converse amongst corporates, NGOs, government departments and other private actors. E-governance provides a wider reach to all stakeholders through a variety of electronic means (Vincent and Harris, 2009: 105). Thirdly, e-governance facilitates the ongoing and unimpeded exchange of ideas, facts and information from both social and technology sources that is free from encumbrance, colour and influence of scale.

informational conduits. E-governance offers the opportunity to provide data more quickly, more widely, and more communally than any other medium. Proponents of e-governance in the complementary sense regard it with greater worth than simply a technical solution, moreover as a proposition that has both a technical and a social architecture (Sassen, 2003b: 3). In e-governance terms the internet presents a system which ties its procedural and social qualities together under seven key features (Malcolm, 2008: 8). E-governance is capable of social and participatory connections that are far richer than other methods. While face-to-face meetings are usually cited as the premium environment for good governance to occur, e-governance outplays other forms in three crucial areas. Firstly, it provides a ubiquitous platform for information sharing at all levels that 'routs' around obstacles such as censorship, corruption and hierarchy. Secondly, the following field study is a noteworthy enterprise, fulfilling all of the major criteria needed for new governance, while illustrating some of the effects that private actors, using electronic mediums, can introduce into Indonesian forestry management. The Landgate Firewatch-Indonesia Program (FWI) uses satellite technology to provide instant reporting on the development of forest fires. In so doing it enables local authorities to act swiftly to minimise the spread of forest fire and to retain large sections of forest intact. It also provides a continuous summary of the scale and pace of the attack on, and change to, Indonesia's forests. It assists to identify the forces and actors that are driving forestry issues, and it does this in real time.[12]

The Firewatch Program

The Landgate Firewatch-Indonesia (FWI) facilitates a 24/7 stream of data. It collects data from satellite imagery and collates the information to create spatial mapping evidence that can identify changes to the landscape, particularly in forested areas. That information is of great use to forestry stakeholders. Moreover it is of particular value because it is delivered in real time. The information is of even greater worth because the Firewatch programme delivers the information across a network environment of open source nodes that makes its distribution seamless, instantaneous and ubiquitous. FWI provides e-governance through three key improvements of vast proportions in relation to previous Indonesian satellite imagery systems. The information is delivered in real time, allowing firefighters a genuine early warning system (EWS) (Dawbin *et al.*, 2009: 1). The data is delivered through a backbone that has dedicated delivery capabilities, thus ensuring that the

12. The term 'real time' refers to the reporting of changes to forest assets as they happen. Previous reporting has always been after the act. It is impossible to repair a tree that has been cut down, or a section of forest that has been burnt out. Real time reporting provides any and all stakeholders in forestry the opportunity to prevent illegal losses on the scale that they have been occurring over the past decade.

information reaches the key stakeholders that it is intended for. Simultaneously the information is also delivered to a parallel open source node that allows ubiquitous distribution of spatial data, changes in the forest landscape, and widespread public awareness (Roswintiarti, 2007: 14).

The Firewatch network is a cleverly crafted set of pathways that has combined the unique MODIS[13] system (NASA, 2008) from the United States and shared land information platform (SLIP) technology from the Australian team at Landgate in order to deliver real time forest-fire data to any and all forestry stakeholders. At one level it is an example of a high technology solution, using cutting edge satellite imagery and state of the art image processing to create a mash up[14] of early warning fire data and existing forest assets.

At another level Firewatch operates amongst local and individual stakeholders at the forest floor. Information is sent from the INHERENT network backbone to universities,[15] who then relay and distribute the information in several ways. The local Indonesian Ministry of Forestry (MoF) offices and outposts are connected with each university using a microwave link. MoF offices are therefore alerted to forest fires in their area in real time so that they can take immediate action. In most cases this includes mobilising MoF officers and coordinating with local services. At the same time, staff and students at universities interpret the satellite information from a local and cultural perspective, using their network of local students to further distribute key early warnings amongst the community, NGOs, and to local forest people of interest. Typically at this level the information is shared in many ways such as: SMS messaging, mobile phones, email, social networks, radio broadcasts, motorcycle courier messages and face-to-face dialogue. Usually it is a mix of all of these approaches.

The two notable characteristics about Firewatch are: a) it works on all levels, and b) information flows back and forth in multiple directions. The information and early warning streams that get to the local outposts are

---

13. MODIS (or moderate resolution imaging spectroradiometer) is the equipment in two NASA satellites orbiting earth. It relays highly accurate information about changes in land (including forestry) for the purpose of predicting global change and protecting the environment.

14. Mash up: the term mash up refers to a new range of web-based applications created by programmers to mix at least two different services from disparate (and even competing) websites. In the Firewatch programme, the opportunity exists for a "mash up" where the Firewatch data is overlayed onto existing maps showing borders and property allocations, wind speed and smoke dispersion models, as well as water data, electricity, roads and a range of infrastructure. The resulting visual can be easily understood by lay people and private stakeholders, and can be distributed using computers and/or mobile phones. The term applies both literally and figuratively.

15. The Indonesian higher education research network (INHERENT) is a high speed (by Indonesian standards) backbone that allows for the low cost rapid transfer of satellite data between LAPAN (Indonesian space agency), the Ministry of Forestry, and end-users.

automated and occur seamlessly to each outpost in a matter of seconds. Simultaneously local university students and staff begin to engage local forest stakeholders on a social level, passing on information and at the same time receiving vital information back in return. In these situations local knowledge combines with spatial knowledge to create a powerful exchange of information that is shared instantaneously, freely and in all directions. Using these combinations of electronic media the constraints of scale and the dilemmas between horizontal and vertical distribution are instantly minimised. Local knowledge is put to great use and becomes integrated with satellite data. On mobile phones and computers, mash ups are superimposing data from local and satellite sources. In a face-to-face context, discussions include the same overlapping themes and ideas. The result is a very fast, yet very rich set of discussions. Local knowledge is critical because understanding conditions at the forest floor can provide for accurate forecasting of potential forestry impacts. Local knowledge about illegal logging activities can be combined with spatial data to predict forest-fire outcomes and expectations. In reality, the combinations are much more detailed because local stakeholders can visualise the impact of their knowledge in a wider sense and therefore contribute to a more elaborate range of local facts.

The Value of E-Governance

The Firewatch example is important because it demonstrates the characteristics of new governance thinking by means of a high-technology e-governance solution. It is a complementary method for improving forestry management, that simultaneously offers hope on a range of forestry issues such as illegal logging, local knowledge, stakeholder participation, forestry protection, and forestry authority. In its own right the Firewatch example is an example of genuine e-governance. Forest management will benefit greatly from engaging with electronic communication, despite these forms of contact ranging from simple SMS messaging to satellite telemetry. What is clear is the speed with which communication is increasing, the increasing range of communication options, and acceptance of a broad range of ICT media. Coupled with the dynamic downward changes in the pricing of all electronics globally, the notion that e-governance will enable forest management to ameliorate its management problems is rapidly improving.

The FWI project provides spatial data and mapping information that is delivered by a variety of electronic and face-to-face means for the purpose of mitigating the effects of forest-fires in Indonesia. The project distributes this information to forestry stakeholders in real time and in a ubiquitous electronic format. The project therefore purports to influence the actions and decisions of local, private, non-government and forest-dependent stakeholders through their understanding, participation, and coexistence in forestry. It is one thing for information to create just-in-time scenarios that reduce forest-fire damage by the timely deployment of resources. However, can that same information-sharing also influence the behaviour of the forest actors who are the primary cause of most forest-fires? The challenge for e-governance is whether it can influence authority and control in the absence of state authority.

Conclusion

It is important to evaluate the influence and impact of e-governance in forest management to ascertain the validity of the claim that electronic mediums are drivers for new modes of governance: and that e-governance forms self-organising networks of participation, authority and cross-boundary accountability. In order for e-governance to hold such a standing in forestry, there must be evidence of self-regulation amongst private actors in forestry, as well as sufficient legitimacy and involvement as a precursor to workable and acceptable enforcement by local authorities. The Landgate Firewatch project does not require to completely eradicate forest fires, or corruption or illegal forestry in order to acquire a legitimate status alongside other vehicles for governance. The diminution of illegal logging and burning, along with the acceptance of non-state-centric authority amongst forestry stakeholders, indicates that e-governance is effective, especially where this involves trans-national agreements or the recognition across boundaries of loose, soft law arrangements and diffuse regulatory standards.

At first glance, a system of providing satellite data as automatic data streams might seem to preclude the need for human participation, and therefore might raise the concerns of those who doubt the ability of e-governance to sit equally alongside other modes of governance. However, to understand e-governance is to understand the importance of a structure that works to complement existing governance rather than to compensate for them. The same arguments about lack of infrastructure, lack of computer understanding, and lack of education are thrown up whenever the notion of electronic mediums in forestry are suggested.

Genuine e-governance develops when the stakeholders develop a shared culture that, in turn, brings to the fore a new set of norms and values. In the case of FWI, the spatial data brings forward a range of other locally assembled informational updates that acknowledge the various indiscretions of illegal loggers, timber companies and local individuals. Knowledge begets knowledge. However, of greater importance is that as the spatial data is repeatedly verified and known to be true, a level of trust builds up amongst stakeholders. Even those stakeholders who may not stand to gain from accurate spatial forestry mapping are still unlikely to challenge a continuous stream of accurate data.

Therefore, when distributed in combination with multiple verifications from the ground forestry locals, the FWI spatial mapping data creates a set of norms and values that laid the foundations for openness, accountability, responsibility and improved forest management.

References

Barr, C. (2001). "Profits on Paper: The Political-Economy of Fiber and Finance on Indonesia's Pulp and Paper Industries", in C. Barr (ed.), *Banking on Sustainability: Structural Adjustment and Forestry Reform in Post-Suharto Indonesia.* Washington DC: CIFOR and WWF.

Bevir, M. (2009). *Key Concepts in Governance.* London: SAGE Publications.

Cheong, L.K. and J.S.L. Yong (2003). "The Future of e-Government- Perspectives and Possibilities", in J.S.L. Yong (ed.), *E-Government in Asia: Enabling Public service Innovation in the 21st Century.* Singapore: Times Editions.

Clarke, J. (2009). "Governance Puzzles", in L. Budd and L. Harris (eds.), *e-Governance: Managing or Governing?.* New York: Routledge.

Corbera, E. (2005). "Bringing Development into Carbon Forestry Markets: Challenges and Outcomes of Small-scale Carbon Forestry Activities in Mexico", in Daniel Murdiyarso and Hety Herawati (eds.), *Carbon Forestry: Who Will benefit?.* Centre for International Forestry Research (CIFOR) at *http://www.cifor.cgiar.org accessed 22nd May 2008.*

Cutler, A.C., V. Haufler and T. Porter (1999). *Private Authority and International Affairs.* New York: Suny Press.

Dawbin, K., R. Smith and R. Couacaud (2009). *Firewatch Indonesia Project—Progress Report April 2009.* Midland: Western Australia Land Information Authority.

Diprose, R. (2007). "Passing on the Challenges or Prescribing Better Management of Diversity? Decentralization, Power Sharing and Conflict Dynamics in Central Sulawesi, Indonesia", Paper presented at the *Conference on Federalism, Decentralization and Conflict.* October 5-7. Retrieved May 11, 2009 from *http://www.crise.ox.ac.uk/copy/decentralization%20conference/diprose.pdf*

Eckardt, S. and A. Shah (2006). "Local Government Organisation and Finance: Indonesia", in *Local Governance in Developing Countries.* Washington D.C.: The World Bank.

FAO (2005). *Needs for Microfinance among Forest-based Small Scale Enterprises,* Food and Agriculture Organisation of the United Nations, Rome, Retrieved September 6th, 2009 from *http://www.fao.org/DOCREP/008/a0226e/a0226e06.htm#Forestbasedsmallscaleenterprises*

Falkner, R. (2003). "Private Environmental Governance and International Relations: Exploring the Links", *Global Environmental Politics* 3(2): 72-87.

Freeman, J. (2000). "The Private Role in Public Governance", *New York University Law Review* 75(3): 543-675.

Gregersen, H. and Contreras-Hermosilla. (2001). "Investing in the Future: The Private Sector and Sustainable Forest Management", in M.E. Chipeta and M. Joshi (eds.), *The Private Sector Speaks: Investing in Sustainable Forest Management.* Jakarta: SMK Grafika Desa Putera. pp. 235-287.

Heritier, A. and S. Eckert (2007). "New Modes of Governance in the Shadow of Hierarchy", *EIU Working Papers,* European University Institute, from the website *http://ideas.repec.org/p/erp/euirsc/p0183.html* accessed 8th May 2008.

Heywood, A. (2002). *Politics* (2nd ed.). New York: Palgrave.

Hill, D.T. and K. Sen (2005). *The Internet in Indonesia's New Democracy* (1st ed.). London: Routledge.

JakartaPost. (2008). "Walhi to up influence on government policies" [Electronic Version]. Jakarta Post, National. Retrieved 20th March 2009 from *http://www.thejakartapost.com/news/2008/04/15/walhi-influence-government-policies.html*

Jaya, W. and H. Dick (2001). "The Latest Crisis of Regional Autonomy in Historical Perspective", in G. Lloyd and S. Smith (eds.), *Indonesia Today: Challenges of History.* Singapore: Institute of South Asian Studies, pp.216-228.

Jessop, B. (1997). "Capitalism and Its Future: Remarks on Regulation, Government and Governance", in *Review of International Political Economy* 4(3): 561-581. Autumn.

Karousakis, K. (2007.) *Incentives to Reduce GHG Emissions from Deforestation: Lessons Learned from Costa Rica and Mexico*. Organisation for Economic Co-operation and Development. *http://www.oecd.org/dataoecd/55/54/38523758.pdf%20accessed%2028th%20May%202008*>Accessed on May 18, 2008.

Kelly, J. (2006). "Central Regulation of English Local Authorities: An example of Meta-Governance?", *Public Administration 84*(3): 603-621.

Kooiman, J. (ed.) (1993). *Modern Governance: New Government-society Interactions*. London: Sage.

Kruseman, G. and L. Pellegrini (2008). "Institutions and Forest Management: A Case Study from Swat, Pakistan", The Fondazione Eni Enrico Mattei Note di Lavoro Series Index, Retrieved September 6th, 2009 from *http://www.feem.it/Feem/Pub/Publications/WPapers/default.htm.*

Latham, R. and S. Sassen (2005). "Digital Formations: Constructing an Object of Study", in R. Latham and S. Sassen (eds.), *Digital Formations: IT and New Architecture in the Global Realm*. Princeton: Princeton University Press.

Legge, J.D. (1961). *Central Authority and Regional Autonomy in Indonesia: A Study in Local Administration, 1950-1960*. Ithaca, N.Y.: Cornell University Press.

Malcolm, J. (2008). *Multi-Stakeholder Governance and the Internet Governance Forum*. Perth: Terminus Press.

McCarthy, J. (2007). "Local Voice in Shifting Modes of Decentralised Resource Control in Central Kalimantan, Indonesia", Australian National University, Research School of Pacific and Asian Studies, Resource Management in Asia-Pacific (*Working Paper* 65).

McCarthy, J.F. and C. Warren (2009). "Communities, Environments and Local Governance", in J.F. McCarthy and C. Warren (eds.), *Community, Environment and Local Governance in Indonesia*. New York: Routledge.

NASA (2008). *About MODIS: The Moderate Resolution Imaging Spectroradiometer*. Retrieved April 20th, 2009, from *http://modis.gsfc.nasa.gov/about/*

Nugraha, A. and B. Rudiantoro (2008). *Neraca Sumber Daya Hutan: Defisit Tengah Keberlimpahan*. Banten: Wana Askara.

Osborne, D. and T. Gaebler (1992). *Reinventing Government*. Reading: MA: Addison-Wesley.

Paetau, M. (2003). "Space and Social Order: The Challenge of Computer-Mediated Social Networks", *Journal of Sociocybernetics* 4(1): 23-35.

Peters, B.G. (1996). *The Future of Governing: Four Emerging Models*. Kansas: University Press of Kansas.

Rametsteiner, E. (2009). Governance Concepts and their Application in Forest Policy Initiatives from Global to Local Levels. *Small-Scale Forestry* 8(2): 143-158.

Ricklefs, M.C. (2001). *A History of Modern Indonesia: Since c.1200* (Third ed.). Hampshire: Palgrave.

Rhodes, R.A.W. (1996). "The New Governance: Governing without Government", in *Political Studies* XLIV: pp. 652-667.

Roswintiarti, O. (2007). *Firewatch Indonesia: Information and Communication Technology Collaboration Plan*. Jakarta: LAPAN. Retrieved 20th April 2009, *http://www.google.com.au/search?hl=enandq=ITB+Firewatchandbtn G=Google+Searchandmeta=andaq=fando*

Sassen, S. (2003a). "Globalisation or Denationalisation?", *Review of International Political Economy* 10(1): 1-22.

———. (2003b). "Digital Formations: Mapping a Field of Inquiry". Paper presented at the IFIP (WG 8.2 - WG 9.4) *IS Perspectives and Challenges in the Context of Globalisation*, Athens; Greece. Retrieved 12th April 2009, from *http://www.aueb.gr/ifip-isglob03/proceedings/Saskia%20Paper.pdf.*

SBS (2002). *The SBS World Guide*. Melbourne: Hardie Grant Publishing.

Speth, J.G. and P.M. Haas (2006). *Global Environmental Governance*, Washington: Island Press.

Steinberg, D.J. (ed.) (1987). *In Search of Southeast Asia: A Modern History*. Revised Edition, Sydney: Allen and Unwin.

Suan, B.H. (2003. "Making e-Governance Happen—A Practitioner's Perspective", in J.S.L. Yong (ed.), *E-Government in Asia: Enabling Public Service Innovation in the 21st Century*. Singapore: Times Editions.

UNECE (2006). "Public Procurement Policies Driving Certification: Certified Forest Products Markets, 2005-2006", *Forest Products Annual Market Review 2005-2006*: Geneva Timber and Forest *Study Paper* 21. Retrieved 23rd April, 2009, from *http://www.unece.org/timber/docs/fpama/2006/fpamr2006.pdf*

UNEP (1999). "Methodological and Technological issues in Technology Transfer: Forestry Sector", [Electronic Version]. *IPCC Special Report*: Intergovernmental Panel on Climate Change. Retrieved 20th May 2009 from *http://www.ipcc.ch/ipccreports/sres/tectran/240.htm*

UNESCO (2005). *Defining e-Governance*. Retrieved 20th March 2009, from *http://portal.unesco.org/ci/en/ev.php-L_DO=DO_TOPICandURL_SECTION= 201.html*

UNF (2009). *United Nations Foundation*. Retrieved March 12th 2009, from *http://www.unfoundation.org/our-solutions/mobile-technology/technology-partnership.html*

Vincent, J. and L. Harris (2009). "Early Adopter" Case Studies, in L. Budd and L. Harris (eds.), *E-Governance: Managing or Governing?*. New York: Routledge.

Visseren-Hamakers, I.J. and P. Glasbergen (2007). "Partnerships in Forest Governance", *Global Environmental Change 17*: 408-419.

Von Furstenberg, G.M. and J. Kirton (2001). "The Challenge Ahead", in G.M. von Furstenberg and J. Kirton (eds.), *New Directions in Global Economic Governance: Managing Globalisation in the Twenty-first Century*. Aldershot: Ashgate Publishing Company.

Walters, W. (2004). "Some Critical Notes on Governance", *Studies in Political Economy* 73: 25-42.

World Bank (2006). *Sustaining Economic Growth, Rural Livelihoods, and Environmental Benefits: Strategic Options for Forest Assistance in Indonesia*.

Wollenberg, E., M. Moeliono and G. Limberg (2009). "Between State and Society: Decentralization in Indonesia", in M. Moeliono, E. Wollenberg and G. Limberg (eds.), *The Decentralization of Forest Governance: Politics, Economics and the Fight for Control of Forests in Indonesian Borneo*. Trowbridge: Cromwell Press.

10

Restoring Indian Forests through Microfinance

K.K. KAUSHAL

A substantial track record of accomplishment and a significant body of empirical studies worldwide together underline the significance of microfinance as an effective antipoverty and development strategy (Khandekar, 2001; Wright, 2000; Zaman, 2000). However, even a well-designed microfinance programme is unlikely to have a positive impact on the poorest unless it specifically seeks to reach them through appropriate product design and targeting (Wright, 2000). It is clear from various reports that there are strong potential synergies between microfinance and the provision of basic social services to clients. However, the services provided need to be relevant to the needs of the target groups and not just an add-on that is of poor quality (Marcus *et al.*, 1999).

Of the 67.7 million hectares (ha) of actual forest area of India (most of which is state-owned) almost 43 per cent is degraded, with a canopy density[1] of less than 40 per cent (FSI, 2007). This degradation of forests is mainly ascribed to the rigid state control and the resulting disempowerment and displacement of indigeneous tribal and hill communities, accompanied by the disintegration of community-based resource management (CBRM) (Fernandes and Menon, 1987; Gadgil and Guha, 1992; Guha, 1991; Kelkar and Nathan, 1991). Consequently the Government of India (GoI) made a major policy shift in 1988 and switched over to joint forest management (JFM). JFM is an evolving policy-based programme, which sets out to establish management 'partnerships' between the local forest dependent communities and the state for the sustainable management and joint benefit-sharing of public forest land (Sarin, 1995). To accomplish this, JFM seeks to shift the existing inequitable distribution of management control by directly involving local people and institutions in forest management (Campbell, 1996). JFM does not involve the transfer of ownership over forests, but instead attempts to restructure the

1. Canopy density indicates the extent to which sunlight is prevented from falling on the ground by tree crowns. Canopy density of 40 per cent and above is taken as dense forest whereas 10-39 per cent is taken as degraded.

formal system of access, decision-making, and sharing of benefits to account for the needs of the local communities. So far, 22 state governments have issued orders to implement JFM and the states have evolved their own mechanisms of involving local communities in conformity with the proclaimed policy. About 36130 village forest councils (VFC) are managing a total of 10.25 million ha of forest area in the country (FSI, 2007). It is a known fact that during the last 140 years of state control most of the land with the potential for agriculture was cleared of forests. The remaining forest in India is now concentrated in rugged and inaccessible areas and this reduces the potential to harvest and manage the forest for production purposes. The abutting population has increased manifold; their needs have also multiplied. It is impossible to create forest-based livelihoods for all the abutting population. Consequently, most of the states are following a two-pronged approach to involve communities; increasing the stake of the neighbouring communities in the management and utilisation of the forests, and creating alternative sources of employment to reduce the pressure on forests (Kumar *et al.*, 2000). But the resources to promote alternative income are limited and cannot wean all forest dependents (Kaushal and Kala, 2004). However, if the limited available funds are used for the creation of a revolving fund for microfinance provision instead of providing doles to a few individuals, then the impact can be larger and more sustainable also.

This chapter seeks to present how microfinance as a component of the JFM project has emerged as a potent tool for the development of forest villages in Tamil Nadu and is reducing people's dependence on forests to non-timber forest products (NTFPs) and indirect benefits. Further it substantiates with field data that the regeneration of forests is directly linked to the successful working of microfinance.

Tamil Nadu Forestry Project

Tamil Nadu, the southern state of India, has a geographic area of 13 million hectares, which constitutes 3.96 per cent of the land area of the country. The total population of the state is 62.4 million (2001 Census) accounting for 6.07 per cent of the country's population. The recorded forest area[2] is 2.29 million hectares, which constitutes 17.6 per cent of the land area of the state. But, half of this actual forest cover[3] is degraded and has a canopy density of less than 40 per cent (FSI, 2007).

2. Forest areas in all lands are statutorily classified as forest, though they may not necessarily bear tree cover.
3. Forest cover refers to all lands with a tree canopy density of more than 10 per cent, though they may not be statutorily classified as forestland.

There are 15,822 villages in the state, of which 1405 are abutting forests. The total population of these forest villages is estimated to be 3.11 million. No separate consumption or income data are available, but it is a known fact that the forest communities are the poorest of the poor (Kaushal and Kala, 2004; Tiwari, 1994). Because of the remoteness and a thin and scattered population, modern amenities have not reached them. Moreover, the development departments hardly reach these areas. Likewise, the cooperative banks and rural banks restrict their operations to the villages in the plains, which have more productive lands that allow their people to make gainful use of loans.

The Tamil Nadu forestry project (TFP) is a JFM (community forestry) programme, funded by the Japan Bank for International Cooperation (JBIC) for Rs. 1,100 crore. This project has been implemented in Tamil Nadu state since 1997-98. In the project, degraded forest micro-water sheds along with abutting habitations are selected. The forest micro-watersheds along with abutting habitations are selected. The forest area is divided into three zones—(i) lowest zone or utility zone, (ii) middle zone or asset creation zone, and (iii) upper zone or eco-restoration zone. Normally, the area for all three zones is 250 ha, in which the zone-wise gap planting is taken up. The unique feature of the project is that it has a provision of Rs. 8 lakh for the development of abutting village population over a period of three years. The aim is to reduce the dependence of villagers on forests by initiating alternate activities for generating income.

In each of the identified management units, the people's representative body, called a VFC, is formed which is fully involved in the planning and execution of works, protection, harvesting and benefit sharing, with focus on degraded forests. One man and one woman from each household, provided that they are willing, are enrolled as its members. Any person who opts out from the membership of VFC is not entitled to any benefits. The VFC meets as and when called for, but in any case at least once in three months.

Each VFC elects an executive committee (EC), with each hamlet electing at least 2 members; each VFC elects a minimum of 5 and a maximum of 15 members to the EC. The *panchayat*[4] members of the management unit are co-opted as ex-officio members of the EC. The members of the EC elect a president from among themselves who is also the president of the VFC. The forest ranger concerned is the member secretary of the EC, who facilitates the election of the members and the president of the EC. The EC is responsible for the day-to-day activities of the VFC and meets at least once a month. A memorandum of understanding (MoU) is signed in the beginning between the

4. The *panchayat* is the lowest tier of local self-government in India.

divisional forest officer (DFO) (on behalf of Forest Department) and the VFC president. This gives the details of the roles and responsibilities of the Forest Department and VFC—the partners in JFM. All the VFCs have been registered under the Tamil Nadu Societies Registration Act, 1975.

The government order sanctioning funds from JBIC soft loan defined the objective of the programme as increasing the tree cover through the involvement of people. But once the programme was launched, it was continuously reviewed, modified and broadened in consonance with the feedback from the people and the field officers. The amount for development of forest dependents was increased to Rs. 8 lakh from Rs. 4 lakh. The individual grants were converted into interest-free loans so that the VFC can build up its corpus fund and extend loans to the remaining forest dependents for acquiring productive assets. The VFC president was made the joint signatory for the withdrawal of funds and the submission of accounts. State government ordered the involvement of all other line departments, like public works department, electricity, health, agriculture, animal husbandry, and tribal development etc., for the holistic development of these micro-watersheds on a priority basis. Thus the project has come to acquire a multidisciplinary approach in which the promotion of microfinance plays the lead role.

TFP and Microfinance

Each VFC opens an account with the nearest post office and bank, which is jointly operated by the president and member secretary. Income from the following sources accrues into that account.

1. Fines imposed for offenses (like grazing or illicit removal) in the JFM area.
2. Recovery of loans.
3. Income from the sale of non-timber forest products[5] (NTFPs).
4. Membership fees.

Out of Rs. 8 lakh meant to be spent on the buffer zone in each village, about 30 per cent is spent on village development and community assets like threshing floors, community halls, etc. The remaining amount is given to individuals for acquiring productive assets. Soon after the launching of the project, this amount was declared as a loan, which individuals have to pay back to the VFC so that it can give loans to other people, as well as later give further assistance to the same individuals. In fact, this buffer zone amount is the major source of VFC fund, as income from the sale of NTFPs will take many more years to be significant.

---

5. NTFPs are goods of biological origin other than wood that are derived from the forests. These include fruits, nuts, tubers, mushrooms, essential oils, medicinal herbs, spices, resins and gums.

Presently, extending credit is the main financial activity of the VFC. But the term microfinance has been applied to the VFC because it arranges insurance for the cattle purchased through its loans. Some VFCs have also taken group insurance policies for their members. Further, NTFP sales are done by the VFC only.

Methodology

Study Area

There are 48 field-level implementing units called forest divisions in Tamil Nadu State Forest Department which have been implementing TFP. Plantation division, Madurai is undertaking TFP implementation in Madurai and Theni districts of Tamil Nadu state. Till 2006, 27 villages have been brought under the project in this division and this study was conducted in those 27 villages of the plantation division, Madurai.

Procedural Description and Results

Data regarding the financial position and recycling of money were collected from the records of each VFC, since each VFC maintains the loan disbursement and loan recovery registers and cashbook. Recovery and recycling were separately rated on a scale of 0 to 1. If loan recovery for the village showed 100 per cent recovery, we gave it a value of 1, and if loan recovery was less, then the decimal value corresponding to the loan recovery percentage was assigned. For analysing recycling, evaluation considered whether the funds after recovery were kept in a bank account or given as fresh loans to new individuals within a reasonable period, with a value from 0 to 1 assigned. We allowed a period of one month and an accumulation of up to Rs. 50,000 for giving fresh loans. As per expectations, the values were almost the same as the better recovery villages were practicing early disbursal. The average of loan recovery and recycling values was taken to arrive at the value of the combined parameter of loan recovery and recycling.

Then a team of three forest range officers (FRO) assessed the protection and regeneration status of each programme area. Better protection through effective social fencing was reflected in the growth of planted seedlings, the regeneration of existing rootstock, and the absence of goat or cattle dung on the forest floor. The team perambulated the forest area of each of the 27 forest villages to assess the protection and regeneration status and independently rated the same on a scale of 0 to 1. The average of the ratings assigned by the three rangers was taken as the value of forest protection status for the respective villages. Data regarding the financial position of each village is presented in Table 10.1, which shows the year of the formation of the VFC, the amount extended as loan, along with the ratings of the VFC fund and forest protection.

Table 10.1

Village Forest Council's Fund Working as of 31.3.2006 in Plantation Division, Madurai, India

Year	Sl. No.	Name of TFP Village	Total Loan Given (Rs.)	Rating of Loan Recovery and Recycling (0 to 1 scale)	Rating Forest Protection Regeneration (0 scale)
1997-98	1	Konapatti	375365	0.40	0.60
	2	Thethoor	344000	0.40	0.60
	3	Krishnapuram and Vasinagar	270400	0.50	0.50
	4	Kaloothu	366000	0.50	0.50
	5	M. Ayyampatti	290249	0.40	0.60
	6	Nallathathunaickanpatti	310000	0.50	0.60
	7	Nagamanaickanpatti	304000	0.40	0.50
1998-99	8	Mondikundu	464480	0.30	0.50
	9	Srirangapuram	495000	0.80	1.00
	10	Thatchapatti	397000	0.50	0.70
	11	Kunnuvarpatti	490000	0.70	0.60
	12	Gunnuthupatti	390000	0.70	0.60
1999-2000	13	Rengarampatti	510000	0.70	1.00
	14	Vannathiparai	516000	0.70	0.80
	15	Chithayagoundanpatti	450000	1.00	0.80
	16	Nottampatti	396900	0.40	0.50
	17	Aruguveli K.S. Puram	440000	0.50	0.80
2000-01	18	Sukkanodai	510000	0.70	0.80
	19	Solaithevanpatti	500000	0.60	0.80
	20	Nehrujinagar	725000	1.00	1.00
	21	Errampatti	210000	0.70	1.00
2001-02	22	Panamooppanpatti	472900	1.00	1.00
	23	Pommampatti	150000	1.00	1.00
	24	Kuranguthoppu	210000	1.00	1.00
	25	Sanampatti	279200	1.00	1.00
2002-03	26	Pandiyarajapuram	287500	1.00	1.00
	27	Sathiravellalapatti Keelamettupatti	200000	1.00	1.00
2003-04		Grand Total	9927794		

The main source of VFC funds is the amount given by the department to the individual beneficiaries for alternative income promotion which they pay back to the VFC concerned. Most of the loans are for purchasing hybrid cattle, because cattle rearing comes naturally to them, and if a village has 20 cattle, the milk procurement society gets started at the nearest road ahead. Hybrid cattle are not amenable to grazing and have to be stall-fed. They yield 10-12 litres of milk per day in comparison to 1-2 litres from indigeneously bred cattle. In some cases artisans have taken loans to improve their profession. In Rengarampatti village, people have taken up iron-file making, and as many as 102 men and women are gainfully engaged in that trade. The forest department has not imposed any alternative income generation activity and the people decide on their own. Almost all the successful VFCs have voluntarily imposed an interest rate of 1 per cent per month on all the loans given by them. For instance, in Nehrujinagar village only Rs. 4 lakh was given from the project fund, but with 1 per cent per month interest, the total amount has grown to Rs. 7.25 lakh. As can be seen in columns 5 and 6 of Table 10.1, there is a direct correspondence between the VFC fund working and forest protection; this shows that once the people get alternate livelihood they need not do woodcutting and consequently forest protection improves. In the villages where loan recovery is poor, the remaining forest dependents could not be provided with productive assets and they continue to indulge in the grazing of goats or the removal of firewood from the forest. In successful villages, the people received loans twice and are therefore able to engage themselves in other vocations. The social fencing of the forest area is complete in such villages and there is no reason why it should not sustain itself in future. The concept of JFM hinges on social fencing, which means the local people agree through their local institutions and mechanisms not to indulge in grazing and woodcutting and to ensure that others also comply with it.

The VFC president handles the loan collection and only in case of default the persuasion by the VFC members and forest staff is adhered to, because most of the people do not have collateral securities and even when they do, the tedious legal process of loan recoveries cannot be pursued by the VFC president or member secretary. Since the forest ranger enjoys a tremendous amount of respect and fear in forest villages, his persuasive role, if enacted sincerely, is enough to keep people paying on their loans even in villages where formal financial institutions have failed.

The results are the poorest in 1997-98 villages, because the project was launched hastily and initially loan recovery was not envisaged. Only in February 1998, it was announced that individual beneficiaries have to pay back the loan amounts to the concerned VFCs. The results are better in recent years due to better interest taken by a fresh set of rangers in the latter years.

Nurturant style of forest ranger is more successful in JFM (Rishi, 2003). Also, the Department personnel have learned the technique of fostering the VFCs in the latter years.

The efficient working of a VFC fund is reflected not only in forest protection but also in the improvement of the living standards of the people, as evidenced through the increase in the number of concrete houses, people switching over to the use of cooking gas from firewood, number of cycles and mopeds, etc.

Conclusions

In JFM programmes, forest regeneration and a profitable yield from NTFPs will take years; hence the people have to be compensated for the lost opportunity cost of grazing and illicit removal to ensure effective social fencing of the forests. If properly applied, microfinance is a potent tool in this regard.

1. In JFM programme villages, forest regeneration and protection show a direct correspondence with the working of VFCs as microfinance institutions.
2. Microfinance is more successful if it is launched as a component of a development package relevant to the people and the area.
3. In forest villages, the Forest Department has more interaction and rapport with the people. Hence any development initiative including microfinance implemented by the forest department has a better chance of success.

The post-project sustainability of the programme is assured in all such villages, where microfinance has been well-established, because the people need not revert to grazing and illicit woodcutting and can pursue alternative livelihoods through microfinance. Also, the VFCs in such villages acquire more popularity and authority, which further helps in forest protection.

References

Campbell, J.Y. (1996). "The Power to Control versus the Need to Use: A Pragmatic View of Joint Forest Management", *Common Property Resource Digest* 37: 9-10. (March).

Fernandes, W., and G. Menon (1987). *Tribal Women and Forest Economy, Deforestation, Exploitation and Status Change*. New Delhi: Indian Social Institute.

Forest Survey of India (FSI)(2007). *State of Forest Report 2005*, Ministry of Environment and Forests, Dehradun, India.

Guha, R. (1991). *The Unquiet Woods: Ecological Change and Peasant Resistance in the Himalaya*. New Delhi: Oxford University Press .

Gadgil, M. and R. Guha (1992). *This Fissured Land: An Ecological History of India*. New Delhi: Oxford University Press.

Kaushal, K.K. and J.C. Kala (2004). "Joint Forest Management in India: Need for Sustainable Livelihood Approach", *International Forestry Review* 6(1): 13-18.

Kelkar, G. and D. Nathan (1991). *Gender and Tribe: Women, Land and Forests in Jharkhand*. New Delhi: Kali for Women.

Khandekar, S. (2001). "Does Microfinance Really Benefit the Poor? Evidence from Bangladesh", Paper presented at Asia and Pacific Forum on *Poverty: Reforming Policies and Institutions for Poverty Reduction*, held by the Asian Development Bank. Manila.

Kumar, N., S. Naresh, A. Yogindra and M. Kinsuk (2000). "India Alleviating Poverty through Forest Development", (*Evaluation country case study series*). Washington, DC: World Bank.

Marcus, R., B. Porter and C. Harper (1999). *Money Matters: Understanding Microfinance*. London: Save the Children.

Rishi, P. (2003). "Leading JFM through Nurturance: An Indian Scenario", *International Forestry Review* 4 (2): 143-148.

Sarin, M. (1995). "Joint Forest Management in India: Achievements and Unaddressed Challenges", *Unasylva* 26(180): 30-36.

Tiwari, D.D. (1994). "Developing and Sustaining Non-Timber Forest Products: Policy Issues and Concerns with Special Reference to India", *Journal of World Forestry Resource Mangement* 7(15): 1-178.

Wright, G.A.N. (2000). *Microfinance Systems: Designing Quality Financial Services for the Poor*. London and New York: Zed Books Ltd., and Dhaka: The University Press Limited.

Zaman, H. (2000). *Assessing the Poverty and Vulnerability Impact of Microcredit in Bangladesh: A Case Study of BRAC*. World Bank. Retrieved from *www.worldbank.org/html/dec/publications/workpapers*

11

Commons, Community Rights and Statutory Laws

Politics of Forest Use in Uttarakhand Himalayas

PAMPA MUKHERJEE

> Forests are all gone now. There used to be thick growth of Sal and Oak when I came to this village as a young bride. I am seventy years now. In fifty years I have seen everything change. I see more guards and officials now rather than trees. The government is there not to improve our condition but to win elections. (Rukmini Devi, Bhansali Village, Nainital)

In many parts of Uttaranchal, the newly created hill state of India,[1] one finds numerous accounts like Rukmini Devi's. For the rural *pahari*[2] community the formation of this new political entity is yet to make a substantive difference to their everyday lives and to the state of forests around them. In fact many of them enumerate incidences of conflict that the villagers have had with the state while asserting their customary claim on forest resources. Few lament that once a composite entity—the forest is now being fragmented and has been demarcated into diverse property regime types like reserved and village *(panchayat)* forests. The existence of multiple property regimes and conflicting rules and regulations has made the relationship between the people, state and forest a contested domain.

Given the above context, the present chapter aims to highlight the dominance of statutory laws over local self-governing institutions and how community space is constricted in the name of participatory governance. It argues that, while for both national government and international donor agencies democratic decentralisation has become a major concern, in implementing their programmes at the field level they tend to treat communities as mere beneficiaries rather than as active partners. It also highlights the fact that the introduction of new joint forest management (JFM)

1. The separate state of Uttaranchal was formed on 8-9 November 2000 as the 27th state of the Indian Union. Earlier it was a part of the larger state of Uttar Pradesh.

2. The term *pahari* implies hill people.

in 1996 has overridden the customary claims of communities on forests practiced over decades. Further, it shows how boundaries and fences have become sites of anxiety, creating artificial enclaves and plots, and in the process excluding communities from their rightful access to the resources. The present study, which is based on extensive field research in the Nainital district of Uttarakhand, draws its insights from in-depth interviews with local residents.

The Field

The region lies in the Central Himalayan zone of the Hindukush Himalayan range and extending from the Gangetic Plain, it rises to around 2,500 metres. Despite the limited availability of arable land in the region, agriculture is the main occupation as majority of the population (78.30 per cent) is in the rural areas. It is subsistence oriented, typified by low productivity and based on coarse grains cultivated under rainfed conditions.

The society is primarily a traditional Hindu society with a distinct *pahari* culture. Although socioeconomic differentiation has increased, village communities are relatively homogeneous compared to high social stratification in the plains. Land distribution is relatively equal with rare cases of land holdings of over two hectares, and landlessness is low. The area's agro-pastoral economy is still predominantly subsistence based with about 50 per cent of rural households, including the rural elite, having high dependence on village commons and forest lands. High male outmigration in search of employment leave the women behind as effective managers of the rural household economy.

Forests in the region are a source of livelihood for rural residents and provide resources such as fodder, fuel wood, green manure, and construction timber. These resources are critical to the household economy. In their absence, effective household incomes would decline substantially. In such a context, determining how forests can be collectively and appropriately managed is vitally important.

Van Panchayats

Van panchayats in Uttarakhand were born out of conflicts and compromises that followed the settlements and reservations of forests in the hills at the turn of the last century. The first government approved *van panchayat* was thus formed in 1921. According to the recent estimates, there are 6,069 *van panchayats* managing 405,426 hectares of forests (13.63 per cent of total forest area) in the state. Most of these have been carved out of civil (protected) forests under the jurisdiction of the Revenue Department. The area under each *van panchayat* ranges from a fraction of a hectare to upto over 2,000 hectares.

It may be mentioned here that the Community forests managed in accordance with Van Panchayat Act is a hybrid of state ownership and community responsibility. In its efforts to manage and control community forest use, forest committees are guided by Revenue Department rules and by the technical advice of the Forest Department. In contrast to civil forests, community forests or *panchayati* forests as they are popularly known are not 'open' forests. Access and use of forests is guided by rules elaborately designed and implemented by the communities. In fact four identifiable working rules exist relating to use, monitor, sanctions and arbitration. Though only notionally or nominally owned by the communities, community forests are in a very real sense common property with an identifiable user group, have finite subtractive benefits and are susceptible to degradation when used beyond a sustainable limit. However what is more important is that the local users consider them as their collective property and in real sense they are not actually divisible. These forests though are not completely immune from misuse and the condition of the forests varies from poor to very good.

Forest Council Act

The Forest Council or the Van Panchayat Act prescribes how *panchayats* (councils) can be formed and impose duties on village *panchayats*. The objective is to protect the forest areas and ensure that the forest products are being distributed among the right holders in an equitable manner. Kumaun Panchayat Forest Rules enacted under the section 28 (2) of the Indian Forest Act, 1927 provides broad guidelines for the supervision and management of forests under the control of *van panchayats*. These forest council rules lay down the broad parameters of management practices to be followed by the *van panchayats*.

The main function of *van panchayats* are as follows:

a) To develop and protect forests by preventing indiscriminate felling of trees and to fell only those which are marked by the Forest Department and are useful from the point of view of silviculture.

b) To ensure that there is no encroachment on *van panchayati* land and that no rules are being violated that are being enacted under Kumaon and Sodic Land Act of 1948 and that no land should be encroached without prior permission for agricultural practices.

c) To construct and fix boundary pillars and to maintain them, 18(c).

d) To carry out the directives of the sub-divisional magistrate in developing and protecting forests, 18(a).

e) To distribute its produce amongst right holders in an equitable manner, 18(e).

f) Twenty per cent of the area of the forest must be closed for grazing every year.

The Punitive Powers

a) They can levy fines upto Rs. 50 with the prior approval of the deputy commissioner (later revised upto Rs. 500).

b) They can seize intruding cattles and impound them under the Cattle-Trespass Act of 1871.

c) They can forfeit the weapons of the offender.

The Administrative and Financial Powers

a) They can sell grass, fallen twigs and stone slates to local people.

b) They can auction trees upto the value of Rs. 5,000 with the approval of the district magistrate and divisional forest officer (DFO). Auction above Rs. 5,000 is done by the Forest Department.

(c) The income realised from resin, timber and fees is distributed as follows: (i) *zilla parishad* is given 20 per cent for creating and maintaining infrastructure, (ii) *gaon sabha* is given 40 per cent for local development schemes if approved by block development committee, and (iii) the remaining is to be ploughed back by the Forest Department for maintenance and development of *panchayat* rules.

The villagers however feel that through the Act, the bureaucracy exercises excessive control over forest *panchayats*. Bureaucrats on the other hand believe that in the absence of Central control, villagers would clear-fell the entire forest (Personal Communication 1994). Nonetheless in analysing the rules it is quite clear that these rules, while making the *panchayats* responsible for proper management of the forests, deny to it necessary authorities which seem to be vested with the revenue and forest officials. For instance in Section 17 of National Forest Management Act, 1976, it is stated that before a watchman or any other paid staff is kept by the *panchayat*, previous approval of the deputy commissioner (DC) is necessary. An offence involving a sum of Rs. 50 can be compounded only with the previous approval of the deputy commissioner. Similarly permission is required if the seized property (stolen timber etc.) is proposed to be sold. Thus the administrative control over the *panchayat* is still with the deputy commissioner, whereas the technical control has been given to the Forest Department.

New *Van Panchayats*

Though the *van panchayats* managed to survive these multiple obstacles and challenges to their authority in many villages, they are faced with two additional threats in the recent years which may further weaken or bring about their demise as relatively democratic and self-governing forest management institutions. Ironically, these new threats have been state initiatives that are presented as 'devolution' policies: the rapid formation of new *van panchayats* under the direction of the Revenue Department; and the introduction of 'village forest joint management' by the Forest Department. Their practical effect has been to transfer still further authority to the state at the expense of the communities.

In the recent years there has been an increase in the demarcation of remaining civil lands as village forests and this has converted the demand driven process into a supply-driven one. Instead of the villagers collectively applying for a village forest, the administration have imposed their decisions, irrespective of the fact whether *panchayats* are required or not as well as ignoring other aspects like ongoing boundary disputes, existing community management arrangements etc. For example in Nainital district itself there were only 61 *van panchayats* at the time of Independence which increased to 495 by 1999.

Further some multi-village *van panchayats* have been reorganised without consulting the villagers, in ways that exacerbated inter-village conflicts. Traditionally, villagers had developed an effective multi-village governance system that was both democratic and equitable. Reorganisation of *van panchayats* has in fact destroyed the traditional culture of resource sharing. Moreover with the division of forest councils neither the forest area nor the species composition could be evenly distributed among the villages. Some are now left with small forest patches with only Chir pine, while others have all the fodder bearing areas. It has also failed to address any of the major problems plaguing the existing *van panchayats*—the lack of effective and easily accessible dispute resolution mechanisms, inter-village inequity in availability of forest areas, erosion of *panchayat* authority and limited control over forest-based livelihoods and income. The rapid formation of *van panchayats*, rather than expanding space for local forest management, seems to be reducing it still further. It is too early to predict the effects of state-driven *van panchayat* formation might have on local livelihoods, social relations, and forest management practices. The process of their formation, however, gives enough reason for concern.

Village Forest Joint Management (VFJM)

VFJM, implemented by the Forest Department, in 1997 has been even more problematic for the villagers. JFM in other states enables villagers to participate in the management of forest lands under the Forest Department's jurisdiction. But in Uttarakhand, the VFJM rules enable the department to become the dominant partner in the management of *van panchayat* and civil forest lands. The land being brought under VFJM falls under the *van panchayats* or the Revenue Department and not the Forest Department's jurisdiction. The latter is managed in collaboration with *gram sabhas*, the democratic institutions of local self-government at the lowest level.

The VFJM rules also provide for forming village forest councils (VFCs) where there are no *van panchayats*. This is an effort to link VFJM with local self-government through *gram sabhas*. These VFCs are expected to be the representative of key local interests, with one seat each designated for women, scheduled castes/tribes, backward castes, and for persons with a particular interest in forests. The *Parthian* (head) of the *gram sabha* is the president of the VFC and the forest guard its member secretary, the two jointly hold the council's accounts. As many *gram sabhas* in the hills cover more than one village, neither the president nor the secretary of the VFC may be the resident of the village whose forest management institution they head. Whereas linking community forest management institutions with those of local government is highly desirable, the order for constituting VFCs is a top-down, mechanical prescription. It says nothing about strengthening participatory governance by the *gram sabhas* and forest committees or their respective roles and responsibilities within the local governance structure. Simply prescribing the *gram sabha* head to also be the VFC's president vests further power and responsibilities in one elected individual, an institutional norm which has already weakened collective decision-making within *van panchayats*. Making a Forest Department functionary the member secretary of a local government committee similarly violates the objective of democratic decentralisation of governance by vesting power and authority in a non-elected representative of the bureaucracy.

Notification of the VFJM rules, together with the issuing of other orders for 'participatory' forestry, was a condition of the World Bank ($65 million loan) for the Uttar Pradesh forestry project over the period 1998-2002. JFM is to receive priority under the project, accounting for about 30 per cent of the total budget. The bank's appraisal document does not provide any analysis supporting the introduction of VFJM instead of strengthening autonomous functioning of existing *van panchayats* in the unique historical context of Uttarakhand. Neither does it specify any process ensuring multiple stakeholder participation in framing the 'participatory' orders. In the event, the responsibility for framing them rests with the Forest Department with no history or experience of working with *van panchayats*. Exclusion of long-standing *van panchayats* from any role in the matter is conspicuous by its absence. *Van panchayats* and other community institutions are treated as the objects of attention, not active participants in redefining their future destiny. The World Bank project simply assumes the desirability of importing the standard JFM model from other states into Uttarakhand, with all its shortcomings, instead of exporting a strengthened *van panchayat* framework to them.

The Impact of VJFM on Communities in Uttarakhand

Though, it is too early to see the impacts of VJFM on forest-based livelihoods and forest quality, the content of the VFJM rules, however, suggests a loss of decision-making space for local villagers. Despite claims to empower local forest users, the rules do much to achieve the opposite. collective control over decision-making.

Prioritisation and selection of villages for VFJM is done by the Forest Department in accordance with several selection criteria, making it a supply, rather than a demand-driven process. 'Spearhead teams' communicate with and develop microplans for selected villages. These teams consist of: one ACF; one ranger or deputy ranger; one forester or forest guard; and two NGO 'social motivators', at least one of whom should be a woman. The social motivators are recruited under contract with NGOs and the teams impart three-weeks training. Experienced *van panchayats* leaders with decades of experience of community forest management have no role as facilitators and technical advisors. Instead they are being motivated to protect forests. The agreement to be signed by the participating villagers refers to them as beneficiaries rather than as equal partners.

In Uttarakhand, NGOs and civil society groups have historically played a strong advocacy role. *Chipko*, for example, was triggered by protests led by the NGO, Dasholi Gram Swarajya Mandal. Today, the NGO movement is split into different camps and factions. The vast majority have been co-opted to work as 'private service providers' for the several donor-funded projects in the region, including the forestry project. Once they have accepted working on project terms, they effectively lose their critical and questioning voice. The overall impact is that today the NGO and civil society movements have been considerably weakened with hardly any concerted public action for protecting people's forest rights.

The World Bank funded forestry project has provided an average of Rs. 20 to 15 lakh for implementing a microplan in each village brought under VFJM. Besides promoting inequity between neighbouring villages, the sudden offer of large sums of money to selected villages with high unemployment and limited opportunities for cash incomes, however, had led to the eruption of major conflicts to gain control over the funds. Even where existing *van panchayats* were functioning well, small groups of elite men, with the least dependence on the forest, had often made alliances with Forest Department field staff to pervert the requirement of obtaining general body resolutions accepting VFJM. The majority of the genuinely forest dependent women and men were neither aware of the content of the VFJM rules nor of the fact that the *van panchayat* rules had become inapplicable. Their main involvement had been reduced to that as of wage labourers.

Conclusions

Despite being an excellent example of state-people partnership which has been relatively successful in managing forest resources in the region, the institutions are facing challenges from unrealistic and target-driven policies which would affect its democratic functioning. There is a need to replicate such institutions in other areas rather than interfering with the existing ones. Moreover NGOs need to play more active role in keeping these institutions alive by bringing the communities to the centre stage of decision-making. In order to strengthen such community-oriented institutions, one needs to identify such similar institutions and undertake comparative studies on the same so that anomalies if any can be removed.

References

Agarwal, C. (1996). "Boundary and Property Rights in Uttarakhand Forests", *Wasteland News*, Society for Promotion of Wasteland Development, New Delhi. February-April.

Agrawal, Arun (2001). "State Formation in Community Spaces? Decentralisation of Control Over Forests in the Kumaon Himalayas, India", *Journal of Asian Studies* 60(1): 1-32.

———. (2003). "Grassland Ecosystems and Agroforestry", *Envis Bulletin* 1(2): 105-111.

Agrawal, R. (1999). "Van Panchayats in Uttarakhand: A Case Study", *Economic and Political Weekly*. September 25.

Arnold, J.E.M. and W.C. Stewart (1991). "Common Property Resource Management in India", *Tropical Forestry Papers* 24. Oxford: Oxford Forestry Institute.

Ballabh, V. and K. Singh (1988). "Van (Forest) Panchayats in Uttar Pradesh Hills: A Critical Analysis", *Research Paper* No.2. Anand, India: Institute of Rural Management.

Bandopadhyay, J. and V. Shiva (1988) "Political Economy of Ecological Movement", *Economic and Political Weekly* 1223-1332. June 11.

Behar, A. and R.K. Bhogal (2000). "People's Social Movements: An Alternative Perspective of Forest Management", Unpublished paper for the CIFOR research project on *Creating Space for Local Forest Management*, Bhopal.

Bhatt, S. (1999). "A Case Study of Some People's Institutions in the Akash Kamini Valley, Garhwal", Unpublished paper for the CIFOR research project on *Creating Space for Local Forest Management*, New Delhi.

Feeny, David (1988). "The Demand for and Supply of Institutional Arrangements", in Vincent Ostrom, David Feeny and H. Picht (eds.), *Rethinking Institutional Analysis and Development*. San Francisco: ICS Press.

Guha, Ramchandra (1988). The *Unquiet Woods: Ecological Change and Peasant Resistance in the Himalaya*. New Delhi: Oxford University Press.

Gururani, S. (2000) "Regimes of Control, Strategies of Access: Politics of Forest Use in Uttarakhand, Himalaya, India.", in A. Agrawal and K. Sivaramakrishnan (eds.), *Agrarian Environments, Resources, Representations, Rule in India*. London and Durham: Duke University Press. pp. 70-190.

Nathan, Dev and Govind Kelkar (2001). "Case for Local Forest Management", *Economic and Political Weekly* XXXVI (30). July 28.

Poffenberger, Mark and B. Mcgean (eds.) (1996). *Village Voices and Forest Choices: Joint Forest Management in India*. Delhi: Oxford University Press.

Sarin, M. (2001). "Disempowerment in the name of Participatory Forestry? Village Forest Joint Management in Uttarakhand", *Forest, Trees and People*, Newsletter 44, April.

Somanathan, E. (1991). "Deforestation, Property Rights and Incentives in Central Himalayas", *Economic and Political Weekly* 26(4). January 26.

Sundar, Nandini (eds.) (2001). *Branching Out: Joint Forest Management in India*. Delhi: Oxford University Press.

12

Industrialisation of the Bamboo Sector

Challenges and Opportunities

AMIR ULLAH KHAN and ARNAB KUMAR HAZRA

Recorded history finds a mention about the various uses of bamboo and cane as early as 7th century AD during the days of King Bhaskara Varma. Yet in India, no quantified information is available regarding the actual demand-supply gap for any of these products except for the paper and pulp industry, and the contribution of bamboo in the family income of various categories of consumers. On the resource front also there is no quantification regarding bamboo availability from non-forest land. Being essentially a forest-based material, bamboo, although a non-wood fibre source is included with wood in all Indian studies, including those pertaining to raw materials for the pulp and paper industry. In the national statistics, therefore it is often difficult to separate the contribution and role of wood and bamboo fibres. On the other hand, non-wood forest products, including bamboo, do not have standard classifications. They are usually classified according to their structure (roots, leaves, bark) or the end use (medicine, food, beverages, utensils, etc.). Bamboo product classification is even more complicated due to its multi-functionality. Most of the economic activities related to bamboo are not recorded officially. Accurate statistical information on trade in Bamboo is also difficult to find, and even internationally. The main reason for this is that international trade statistics suffer from outdated customs codes.

Bamboo: Its Characteristics and Uses

Many characteristics of bamboo make it an enduring, versatile and highly renewable resource. Bamboo has more than 1,500 documented uses, ranging from fuel wood to light bulbs, medicine, poison and toys to aircraft manufacturing. Over 1,000 million people live in houses made of bamboo or with bamboo as the key structural, cladding or roofing element.

- It's biological characteristics makes it a perfect tool for reducing carbon dioxide levels in the atmosphere. It generates more oxygen than equivalent strands of trees, lowers light intensity, protects against ultraviolet rays and is an atmospheric and soil purifier.

- Bamboo is an enduring and versatile natural resource. The great diversity of species makes bamboo adaptable to many environments.[1] It tolerates extreme precipitation from 30 to 250 inches of annual rainfall. A dense bamboo cover also offers stakes to trees, fodder to animals and food to humans.
- Bamboo grows very fast and has a short growth cycle. Bamboo not only grows much faster than wood, it also needs relatively little water. It is the fastest growing canopy, growing three times faster than most eucalyptus species. Commercially important species usually mature in 4 to 5 years (*versus* 10 to 25 years for most soft woods). Annual harvests are subsequently possible.
- Bamboo prevents soil erosion. Its anti-erosion properties create an effective watershed, stitching the soil together along fragile river banks, deforested areas, and in places prone to earthquakes and mud slides. The sum of stem flow rate and canopy intercept of bamboo is 25 per cent which means that bamboo greatly reduces rain run-off, preventing massive soil erosion. Thus, bamboos help control landslides, keep flooded rivers along their natural course and slow the speed of the water flow.
- Bamboo is foremost in biomass production, with up to 40 tonnes per hectare per year in terms of culms only in managed stands. An estimated one-quarter of the biomass in tropical regions and one-fifth in subtropical regions comes from bamboo.
- Bamboo has been used in ancient medicine. Bamboo has for centuries been used in *Ayurveda* (for example, *Chyawanprash*) and Chinese acupuncture. The powdered hardened secretion from bamboo is used internally to treat asthma, cough and can be used as an aphrodisiac. In China, ingredients from the root of the black bamboo help in treating kidney diseases.
- Bamboo is one of the world's best natural engineering materials. Due to its high tensile strength, it is an essential structural material in earthquake architecture and is one of the strongest building materials. Its strength-to-weight ratio is better than that of teak wood and mild steel. Bamboo's tensile strength is 28,000 lb per square inch *versus* 23,000 for mild steel. This makes bamboo wood a potential

---

1. It provided the first re-greening in Hiroshima after the atomic bomb blasts in 1945.

alternative, at least in some applications, to steel which requires more energy for manufacturing/production. Its strength and flexibility make it a viable material for building shelters that offer protection against hurricanes and earthquakes.[2] In Bangladesh, 73 per cent of the population live in bamboo houses. Also bamboo-based prefabricated houses can be constructed quickly with new and emerging techniques and is thus an important post-disaster relief material. It is extensively being used in Tsunami rehabilitation in India. Bamboo reinforcement in concrete piles is used by the Indian Railways.

- As a food source, bamboo shoots have provided nutrition for millions of people worldwide. In Japan, the antioxidant properties of pulverised bamboo bark prevents bacterial growth and it is used as a natural food preservative. Taiwan alone consumes 80,000 tonnes of bamboo shoots annually constituting a $50 million industry.
- Bamboo is a viable replacement for wood. Its qualities of strength, light weight and flexibility make it a viable alternative to tropical timber that is used in the furniture and building materials industries.
- It is a critical element of the economy. Bamboo and its related industries provide income, food and housing to over 2.2 billion people worldwide. There is a 3-5 year return on investment for a new bamboo plantation.
- Bamboo is a renewable resource for agro-forestry products. Bamboo is a high-yield renewable natural resource. Ply bamboo is now being used for wall panelling, floor tiles, for papermaking, briquettes for fuel, raw material for housing construction, and rebar for reinforced concrete beams. It can be used to produce many items of daily use that are currently made out of plastic or other less eco-friendly materials.
- Bamboo is being used as an input or raw material in certain industries. It has primarily been used in the paper industry in bulk quantities as a raw material for paper pulp. Bamboo is also used in manufacturing wood substitutes, composites, utility products including Venetian *agarbatti* (incense sticks).
- Bamboo is also a source of energy. Gasifiers can produce electricity using bamboo as fuel. These can also be used for thermal applications replacing furnace and diesel oil. Charcoal and its processed form in powder and briquettes can also be manufactured. It is superior to

2. In Limon, Costa Rica, only the bamboo houses from the National Bamboo Project stood after a violent earthquake in 1992.

other sources of charcoal in terms of calorific value. Bamboo charcoal can also be used as a raw material for activated carbon manufacturing which is used as adsorbent in different industries like vegetable oil, beverage, pharmaceuticals etc. Goldsmiths prefer bamboo charcoal for making jewellery.

In sum, bamboo's excellent growth, environmental, mechanical and engineering properties make it a fine alternative to tropical timber. Its potential for different value added products and application makes it an extremely important material for dispersed employment generation and economic activities. Perhaps these properties and potential usage coupled with increased urgency of environmental issues, have been sufficient to change the attitude towards bamboo, and helped to solve the problems of tropical deforestation. However, Bamboo is still dubbed the 'poor man's timber', representing a social stigma.

Bamboo-based Products: An Appraisal

Literature regarding the multiple uses of bamboo highlights the utility of bamboo for house construction, bamboo ply, agricultural implements, handicraft, irrigation, brooms, medicine, food, fuel, fodder, paper and pulp etc., especially bamboo as a perfect substitute for some wood-based products. The products that can be made from Bamboo can be broadly be categorised into:

1. wood substitutes and composites,
2. industrial use and products,
3. food products, and
4. construction and structural applications.

Apart from this broad classification various handicraft and cottage industry products are also made from bamboo. However, this category of products is not discussed as bamboo-based industrialisation and its prospects limit the scope of the study. Also the input of bamboo as a resource raw material in the paper and pulp industry is also not explicitly dealt with as a bamboo-based product but discussed in the next chapter under the resource situation in India.

Wood Substitutes and Composites

This category of products essentially comprises boards and sticks of varying descriptions and uses, which can further be used to manufacture finished products like wooden floors or blinds or goes into another industry as an input like incense sticks.

Bamboo-based panels: China started producing bamboo panels in the early 19th century. At present more than 20 different types of panels are produced in Asia. Bamboo fibre is longer than wood fibre, which gives bamboo some technological advantages. The panels are widely used in modern construction as structural elements or as forms for concrete mouldings. They are also used for flooring, roofing, partitions, doors and window frames. Bamboo panels have some advantages over wooden board due to their rigidity and durability. Various types of bamboo veneers, panels and boards can be broadly classified as follows: veneers, strip boards, mat boards, fibreboards, particle boards, medium density boards, combinations of these, and combinations of these with wood and other lignocellulose materials and inorganic substances. Composites of bamboo and jute are also possible to make panels.

Bamboo flooring: Bamboo flooring is a quality product that can be used widely and has a large, global consumer market. It has certain advantages over wooden floors due to its smoothness, brightness, stability, high resistance, insulation qualities and flexibility. Bamboo flooring has a soft natural lustre and maintains the natural gloss and elegance of bamboo fibre. This flooring is attractive to the demanding markets in Europe, Japan and North America. The estimated annual production of bamboo flooring in China was 17.5 million square metres in 2004, with about 65 per cent being exported (Customs General Administration of China, 2004).

Bamboo sticks for blinds and incense industry: The art of making screens and blinds from bamboo is not new to India. For centuries, people have woven elegant screens from bamboo that have provided privacy, protection from the sun and added aesthetic appeal to living spaces. Mechanised blind making units can be economically viable enterprises. Again, bamboo sticks making units can substitute the wood that is used in the incense stick, and that industry in India is estimated to be worth US $400 million. It can also be used in match sticks.

Bamboo furniture: Traditional bamboo furniture uses natural round or split bamboo. A new type of 'pack-flat,' 'knockdown' furniture uses glue-laminated bamboo panels. Unlike the traditional design, this furniture may be shipped in compact flat packs, to be assembled on the spot. The new design overcomes many of the problems of traditional bamboo furniture, such as high labour and transportation costs, low productivity, instability, varying quality and susceptibility to insects and fungi. At the same time, it retains the distinct physical, mechanical, chemical, environmental and aesthetic features of bamboo. Export of laminated bamboo furniture is growing rapidly. However, trade statistics currently do not capture the value, owing to the absence of a special code for bamboo furniture. It is usually classified as wooden furniture.

Industrial Products

Traditionally the industrial use of bamboo has been in the paper and pulp industry. Apart from this, the industrial products from Bamboo, essentially comprises converting into fuel or electricity through gasification. Through pyrolysis, bamboo can be converted into three valuable products—bamboo charcoal, oil and gas. Changing the pyrolysis parameters can change the product shares depending on the purpose and market conditions. Bamboo-based producer gases can be used as a substitute for petroleum. Bamboo charcoal is an excellent fuel for cooking and barbequing. There can also be the use of activated charcoal. This is used as a deodorant, purifier, disinfectant, medicine, agricultural chemical and absorbent of pollution and excessive moisture. The industries can use bamboo waste for gasification and thereby produce electricity.

Bamboo for paper and pulp: Several bamboo-producing countries, such as China and India, use bamboo in paper and pulp. Bamboo paper has practically the same quality as paper made from wood. Its brightness and optical properties remain stable, while those of paper made from wood may deteriorate over time. The morphological characteristics of bamboo fibres yield paper with a high tear index, similar to that of hardwood paper. The tensile stiffness is somewhat lower compared with soft wood paper. The strain strength is between that of hardwood and softwood papers. The quality of paper may be improved by refining the pulp.

Bamboo charcoal for fuel: Bamboo charcoal is traditionally used as a substitute for wood charcoal or mineral coal. It can serve as a fuel, absorbent and conductor. The calorific value of bamboo charcoal is almost half that of oil of the same weight. Activated bamboo charcoal can be used for cleaning the environment, absorbing excess moisture and producing medicines. The absorption capacity of bamboo charcoal is six times that of wood charcoal of the same weight. China is a leader in its production. At present, Japan, the Republic of Korea and Taiwan province of China are the main consumers, but its importation is rapidly expanding in Europe and North America. There are three main reasons contributing to the success of bamboo charcoal in international trade:

- bamboo grows faster and has a shorter rotation compared with tree species;
- the calorific value and absorption properties of bamboo charcoal are similar to or better than those of wood charcoal; and
- it is cheaper and easier to produce.

Bamboo-based gasifier for electricity: Gasification of bamboo can produce energy and a range of valuable byproducts. It reinforces a commitment to clean and renewable electricity and thermal energy. It can utilise waste generated by processing operations, substitute the use of fossil fuels, and lower operating costs. Bamboo can be cut into small pieces and used in the gasifier. The requirements for the gasification units are a small proportion of the total availability. A 100 kw gasifier would require only about 1000 tonnes per annum, the equivalent of a truckload every three days on an average. An added advantage of gasification of bamboo is that 15 per cent of the biomass would also be available as a byproduct in the form of high grade charcoal. In the case of a 100 kw gasifier, around 135 tonnes of charcoal would be available each year to meet local needs of fuel. It is clean, cheap and a renewable source of energy. Further, it does not depend on quality, species, and maturity of bamboo.

Bamboo-based fibre and fabric: The most recent advancement in bamboo is the manufacturing of fibre for making yarn and weaving it into various fabrics. There are several spinning mills using. 100 per cent bamboo yarn, and Indian companies such as Raymond, BSL Ltd. of Bilwara group and Paramount Textile Mills Ltd., Madurai, have already launched fabrics made out of bamboo. Bamboo fabrics are naturally anti-microbial and due to the presence of micro-pores in the fabric, they absorb three times more moisture than cotton, making it a superior product.

Apart from the ones outlined above, bamboo extracts contain valuable elements that can also be used as input in several industrial products. For example, bamboo can be used in pharmaceuticals, creams and beverages. Traditional medicines like *Chyawanprash* use bamboo extracts.

Food Products

Under this category, it is essentially bamboo shoots that are consumed after being cooked. Bamboo shoots carry the potential of value added economic activity at the entrepreneurial and community level through cultivation, processing and packaging. Its use in food and cooking goes far back in history. China earns US $130 million annually from exports of edible bamboo shoots. About 200 species of bamboo can provide edible and palatable bamboo shoots.[3] Fresh bamboo shoots are delicious and healthy, with high fibre content. Bamboo vegetables can be found in Chinese grocery stores and restaurants worldwide. After cooking, the shoots are still crisp, because

---

3. Within monopodial bamboos, the main species are Acidosasa edulis, Chimonobambusa quadrangularis, Phyllostachys heterocycla var. pubescens, P. praecox, P. dulcis, P. iridescens, P. makinoi, P. nuda, P. prominens, P. sulphurea cv. viridis, P. vivax, Pleioblastus amarus and Qiongzhuea tumidissinoda. Within sympodial bamboos, the main species are Bambusa rigida, B. pervariabilis, Dendrocalamus latiflorus, D. asper, D. brandisii, D. hamiltonii, Dendrocalamopsis oldhami, D .beecheyana, D. beecheyana var. pubescens, D. stenoaurita, D. vario-striata and Schizostachyum funghomii.

cooking does not destroy their texture. Cooked bamboo shoots can be stored in containers and shipped worldwide.

Construction and Structural Applications

Advances in structural engineering and the development of bamboo composites have opened new vistas for lightweight, durable and aesthetic construction for a variety of applications, enabling informed choices for housing, community and functional structures.

Bamboo housing: There are three main types of bamboo housing: (a) traditional houses, which use bamboo culms as a primary building material; (b) traditional bahareque bamboo houses, in which a bamboo frame is plastered with cement or clay; and (c) modern prefabricated houses made of bamboo laminated boards, veneers and panels. These buildings are usually cheaper than wooden houses, light, strong and earthquake resistant, unlike brick or cement constructions. New types of prefabricated houses made of engineered bamboo have distinct advantages. They can be packed flat and transported at a reasonable cost. They are better designed and environmentally friendly. Bamboo materials are widely available and can be cultivated at a low cost.

The properties and uses of bamboo vary at different stages of growth and its appropriateness at different stages of growth is as follows:

Table 12.1

The Properties and Uses of Bamboo

Up to 30 days	Bamboo shoots to be used as food
Between 6-9 months	For basketry
Between 2-3 years	For laminates and boards
Between 3-6 years	For construction

Note: Bamboo gradually loses strength after the sixth year and up to 12 years.

Bamboo-based Industrialisation: Prospects and Problems

This natural resource has played a major role in the livelihood of rural people and in rural industry, especially in tropical regions. Over 2.2 billion people world over are dependent on bamboo and its related industries for income, food, and housing. Although the rural communities have traditionally been using bamboo, the utilisation has been highly localised as bamboo has often been viewed as an inferior substitute for timber. For example, over 1 billion people in the world live in bamboo houses, yet there has been little effort to build such houses (using prefabricated structures or otherwise) commercially. Traditionally, bamboo has been harvested in the natural forest and its use has been limited to temporal constructions and low-quality utensils prone to rapid decay. Consumption or utilisation has therefore been direct and restricted to poor people with low income and low purchasing power. Market linkage has as a consequence been weak or non-existent in most countries including India.

Yet, bamboo, as noted, has versatile uses as building material, paper pulp resource, scaffolding, agriculture implements, weaving material, plywood and particle board manufacture, basketry, furniture, pickled or stewed bamboo shoots, medicines, etc. Resource management and technical improvements can convert this fast-growing grass into a durable raw material for construction purposes and a wide range of semi-industrialised products.

New industrial applications and modern construction design have both demonstrated bamboo's huge potential, but the bamboo sector in China is the only one reported to be thriving. The Chinese has been able to successfully industrialise the use of bamboo by integrating the bamboo sector with domestic and international markets. In the last 20 years, China has established an integrated chain of bamboo plantations, its semi-processing and industrial product manufacturing such as bamboo flooring, furniture, furnishings, charcoal and fresh bamboo shoots for the domestic and export markets. Its focused intervention to harness bamboo's potential has led to increase in its productivity by more than 10 times since 1970 when it was 2-3 tonnes/ha/annum.

The combined value of internal and commercial uses of bamboo in the world is about Rs. 50,000 crore annually. This is supposed to double by 2015. More than half of the world's consumption of bamboo is in China. China's export of bamboo products is close to Rs. 10,000 crore. As against this, India's size of the domestic bamboo economy is estimated at Rs. 2,043 crore. The market potential was, however, estimated at Rs. 4,463 crore, which could grow to Rs. 26,000 crore by 2015 (Planning Commission, 2003).

The bamboo sector in most other countries is still a part of the informal and backward rural economy. There has been an inability to grab the large potential, which has been successfully demonstrated by the Chinese bamboo industry. This raises the question of the bottlenecks facing bamboo development. Many of these inhibiting factors are at the policy level and are adding to a lack of knowledge among the important stakeholders and a widespread stigma of bamboo as a poor man's timber. In India it has been no different. Both the law and lack of awareness about its industrial application has been the primary bottleneck inhibiting a bamboo based industrialisation process from taking shape. Presently it is underutilised and found in abundance.

The biggest impediment for the development of bamboo based sector has been the irregular and scant supply of bamboo for entrepreneurial use. The paper and pulp industry in India, which has been traditionally using bamboo for over half a century, has constantly innovated to reduce the use of bamboo in its manufacturing process due to this uneven and scant supply. And after the consumption of the paper mills (who usually have long-term contracts with the Forest Departments), very little is left for any other application. This pattern is true for all Indian states. The present regulatory regime in India is the unambiguous culprit for this irregular and inadequate supply.

An efficient regulatory institution is essential for markets to grow in a sustainable manner, especially where environment concerns are coupled with business development. Transaction costs must be minimal, information availability maximal with a clear focus on maintaining the forest cover. Unfortunately, the regulatory structure as regards the bamboo industry has remained caught in the quagmire of archaic forest laws, whereby bamboo is defined to be a tree, and therefore felled bamboo is classified as timber which is subject to transit and trade restrictions. Bamboo is also subject to harvesting permissions in many parts of the country if grown on private lands and which then becomes the basis for imposing the need for transit permits. This has resulted in throttling of the bamboo sector and has discouraged private plantations. The irregular and scant supply of bamboos for processing, despite the world's largest area under bamboos has been a natural corollary.

Clearly the expansion of a bamboo-based sector has not happened due to the restrictions in place. If the restrictions are removed, the sector still might not grow, but can impact livelihood benefits percolating down. This should justify an initial policy initiative through subsidies, incentives and other handholding measures. Economic subsidy can be justified when social benefits outweigh private benefits. So in the bamboo sector the understanding of livelihood benefits is crucial.

The economic and social benefits for example, from activities related to bamboo-based value added products and applications was worked out to be 8.6 million jobs (new) in the Tenth Plan, besides building up large bamboo resource and market opportunities worth Rs. 6,500 crore with an investment of Rs. 2,600 crore, enabling 5 million families of artisans and farmers crossing the poverty line, according to the National Bamboo Mission.[4] The expansion of handicraft, cottage and tiny sector can potentially create 3 million jobs, according to estimates of the Planning Commission (2003). On the other hand, generation of power through gasifiers using bamboo resources exemplifies

---

4. See *http://agricoop.nic.in/bamboo/bamboomission.htm*

assiduous application of technology that can alleviate the present power shortage in most states and thus help improve the overall economy.

There has been a growing awareness in recent years about the importance of bamboo being an important means of economic growth and of improving the socioeconomic conditions of the rural poor. Bamboo as an industrial material can substitute wood to a great extent and that too at a low cost. Bamboo has been traditionally harvested from forest lands in India and the homesteads which may have a few clumps of one of the many species of bamboo for household use but very little intervention in terms of purposive planting has been done in the past. Convincing and informing users and policymakers of bamboo's versatility may fit in with a strategy of poverty alleviation and reducing pressure on tropical forests. Smallholders at the forest fringe can, in particular, improve their livelihood by processing bamboo or growing it in their backyard. Bamboo as a resource needs to be seen as a form of development, with the primary value addition done closer to the resource in order to reap the livelihood benefits. At the same time, a large stock of bamboo contributes to broader environmental goals of erosion control, reforestation and watershed management.

In India, bamboo is mostly found in the forests. As per Forest Survey of India (1999) estimates, 9.6 million hectares forest area of the country contains bamboo amounting to 12.8 per cent of the forest cover. India has the largest area under bamboo in the world, which is estimated to be around 11.36 million hectares. India is also very rich in bamboo diversity. It is the second richest country in the world in terms of genetic resources, after China.[5] Sharma (1988) reported 136 species of bamboos, across 22 genera, occurring in India. Out of these, 19 are indigeneous and 3 are exotic. Naithani (1993) reported 124 indigeneous and exotic species, across 23 genera, to be found naturally and/or under cultivation in India.

The distribution is, however, not uniform. The rich areas are confined to the northeastern parts of the country, Siwalik Hills of Uttar Pradesh, Bastar, region of Madhya Pradesh, Western Ghats in South India and the Andaman Islands. The northeast is the richest source. Fifty-eight species of bamboo belonging to 10 genera are distributed in the northeastern states alone. Around two-thirds of the growing stock or 66 per cent of the growing stock of bamboo in India is found in the northeastern states, but with just 28 per cent of the total area under bamboo in the country. Madhya Pradesh has the second highest area under bamboo, estimated at 20.3 per cent of the area and with 12 per cent of the growing stock. The details of the bamboo growing areas (in forests) and growing stock of major states is given in Table 12.2.

---

5. China with 300 species is leading in genetic diversity of bamboo.

Table 12.2

Major Regions/States by Area under Bamboo

(Percentage)

State/Region	*Area (Percentage)*	*Growing Stock (Percentage)*
Northeast	28.0	66
Mizoram	8.45	13.18
Assam	7.54	16.23
Arunachal Pradesh	4.21	11.91
Manipur	3.39	13.88
Meghalaya	2.89	5.34
Tripura	0.86	1.04
Nagaland	0.70	4.43
Madhya Pradesh	20.3	12
Maharashtra	9.9	5
Orissa	8.7	7
Andhra Pradesh	7.4	2
Karnataka	5.5	3
Others	20.2	5

Source: Madhab (2003) and the National Horticulture Mission at *(http://agricoop.nic.in/AgriMinConf/National_hort_Mission.ppt.)*

Clearly from the table above, states like Manipur and Arunachal Pradesh, within the northeast and Orissa otherwise, have much higher productivity than rest of the states.

Productivity in India

The annual yield of bamboo per hectare varies around 2 tonnes per hectare per annum, depending on the intensity of stocking and biotic interferences. It is however, known that yield in rainfed areas can increase 4 to 5 times in five years if protection from grazing is ensured and proper management practices (soil working, fertilisation and thinning) are adopted (Lakshmana, 1994). The yield per hectare is very low compared to other countries such as Japan, China, Taiwan and Malaysia. As compared to China and Taiwan, India's productivity is one-fourth to one-fifth. India has a long way to go on scientific cultivation of bamboo. Unfortunately, bamboo has no parent in the governmental setup. Only in the homesteads, farmers take care of bamboo. But the lack of scientific approach to cultivation prevents higher productivity.

With regard to the potential productivity of bamboo from forest areas in India, two observations can be made. Firstly, the present actual productivity is far less than the potential and in either case less than what is noticed in the homesteads. Moreover, even in the homesteads, bamboo cultivation is seldom accorded the attention and silvicultural requirements, which would come in when bamboo plantations are undertaken.

As per the Orissa Bamboo Policy, the potential productivity of bamboo forest areas in Orissa is estimated to be 250,000 metric tonnes per annum, but the actual productivity has been around 70,000 metric tonnes per annum, much below the potential. For Maharashtra, the potential productivity of bamboo forest areas have been estimated in the Draft Maharashtra Bamboo Policy to be 1,200,000 metric tonnes per annum, but actual productivity over last few years has been around 200,000 metric tonnes per annum. On the other hand, it is estimated that about 50,000 tonnes of bamboo is extracted from homestead areas in Orissa, with an average productivity of 5-6 tonnes per hectare per annum, much above the productivity level of bamboo in forest areas. In fact, the Orissa Bamboo Policy acknowledges that there is a need to enhance the productivity up to 20–25 metric tonnes per hectare per annum through scientific cultivation, superior clones and species change. Correspondingly, in Maharashtra, around 25,000 tonnes are extracted from homestead areas, with an average productivity of 3-4 tonnes per hectare per annum, which also is much above the productivity level of bamboo in the forest areas. Similarly, the current yield in Andhra Pradesh is 7 tonnes per hectare per annum from forest plantations, while in the managed plantations it was found to be 12 tonnes per hectare per annum.

Traditionally the Forest Department's bamboo harvesting policy systematically maximises dry bamboo output for paper mills rather than green bamboo output for artisans and mature bamboos (2-4 years old) for the industrial needs (apart from the paper industry). In fact, if bamboo forests are carefully worked and green bamboos/mature bamboos regularly harvested, bamboo output of an average clump would jump. Till date the management of state bamboo resources has many constraints with lack of post harvest treatment and technology for product development, inadequate trained manpower and inadequate infrastructure for large scale harvesting in the event of gregarious flowering.

In India bamboo has been primarily grown in forests, which are government-owned land. The exceptions, as noted earlier, were Nagaland and Kerala. Although it is a well established fact that bamboo in India is primarily harvested for supplying to the paper and pulp industry as a raw material, and otherwise used by the rural communities for self-consumption, it is very difficult to obtain even rough estimates of the consumption or utilisation pattern. Tewari (1992) has done the only estimate on consumption pattern and this is given in the Table 12.3. There are two points to note in his consumption pattern outlined. First, the estimates of Tewari (1992) do not include prefabricated houses using bamboo. In the housing sector, bamboo is used in different ways as a building material for roof structure in form of purlins, scaffolding, rafters, reapers, as reinforcement in foundations and in mud walls, flooring, doors/windows, walling, ceiling, water storage tanks, man-hole covers and even for roads in slushy areas. The other point to note is that the estimated percentage used in the paper and pulp industry was 35 per cent in 1992. This fact looks quite circumspect.

Table 12.3

Consumption Pattern of Bamboos in India

(Percentage)

Uses	*Consumption*
Pulp	35
Housing	20
Non-residential	5
Rural uses	20
Fuel (non-industrial)	8.5
Packing, including basket	5
Wood-based industries and Transport	2.5
Furniture	1
Others, including ladders, mats etc.	3
Total	**100**

Source: Tewari (1992).

Bamboo as a resource in India has been used as an input for the paper and pulp industry. For example, in Orissa, around 98 per cent of the bamboo extracted from forests is supplied to the paper mills.[6] The balance quantity is utilised by the local communities for their livelihood and utility products. Of the estimated 50,000 tonnes/annum extracted from homestead areas, around 30 per cent is supplied to the paper mills and balance is utilised for construction, fencing, crafts, agriculture implements and other utility products. This scenario is not much different in other states. In Maharashtra, around 85 per cent of the bamboo extracted is supplied to the paper mills and the balance quantity is utilised by the local communities.[7]

In Assam, the overwhelming industrial use for bamboo is still for pulp and paper. The paper mills in the state have a capacity of 800,000 tonnes per annum, met largely from Assam, but to a lesser extent from the neighbouring

6. Draft Orissa Bamboo Policy, Annexure I.
7. Draft Maharashtra Bamboo Policy, Annexure I.

states. Much of the bamboo utilised in these spheres comes from the forests through a system of contracts, leases and departmental operations. According to a survey report of the state Forest Department of Jharkhand for example, 75 per cent of bamboo is used for pulp and paper, 23 per cent for household and constructional needs, and 2 per cent for bamboo-based cottage industries. Similarly in Andhra Pradesh, the growing stock of bamboo is 3.8 million tonnes, and about 2 lakhs metric tonnes are removed annually. Of this, around 1 lakh tonnes or 50 per cent is supplied to the three paper mills in the state and the remaining is made available to domestic sectors and *Burood* societies.

As noted before, the estimated annual harvest of bamboo in India is 13.47 million tonnes against the current domestic demand of 26.69 million tonnes. Therefore there is already a substantial shortfall in the supply of bamboo in India.

Bridging the Supply Gap

Given that the demand for bamboo as a resource for the paper and pulp industry as well as for the industrialisation needs of the sector is outstripping supply and will continue to do so, there are three things that need to be done. The first is to embark on a systematic plan to rejuvenate the degraded bamboo forest areas by the Forest Department. The second measure would be to increase the yield. Therefore, better management and harvesting techniques need to be adopted along with a systematic working plan for bringing in degraded forest lands under bamboo cultivation. The present system of using the paper mills to do the harvesting is not sustainable. We need a system of 'selective felling' wherein the more matured bamboos are cut and the younger ones are left untouched. Since bamboo in India grows in clumps, and some of them are of the thorny variety (like in Kerala), the method of selective felling that is used is like a 'horse shoe'.

Third, however, since the bamboo from state forests continue to be contractual to the paper and pulp mills, promotion of private plantations of bamboo is essential, especially for meeting the industrialisation needs of the bamboo-based products and applications. This can be a prioritised activity for the Forest Department and giving a thrust to bamboo resource management as a part of natural resources management activities, by promoting seedling distribution and plantation activities to regenerate the replenished bamboo resources. This process can also involve the local communities, for that can have desirable outcome as has been demonstrated in the Andhra Pradesh Community Forest Management (APCFM) project (with the help of World Bank). About 50,000 hectare of degraded bamboo forests were targeted for treatment and the target was exceeded by more than 5,000 hectare (or by more than 10 per cent) by 2005. This can help in the development of site-specific forest inventory, forest management, and integration with working plan, and with the help of other related technical practices including adaptive research and development, can enable the bamboo-based industries to emerge as a viable livelihood enhancement and poverty reduction strategy.

The total forest cover in India, as per the Forest Survey of India, 1999 report was estimated to be 63.73 million hectare, out of which 25.51 million hectare is degraded forest and another 5.19 million hectare is scrub. Thus, total degraded forest is thus about 30.70 million hectare. Thus substantial addition to the existing bamboo stock can be achieved, especially through concerted effort by the Forest Department. The average productivity of bamboo from forests is around 5 tonnes per hectare per annum or less, and which can easily be increased by at least five times, to 25 tonnes per hectare per annum or more. Increasing productivity requires proper management of the bamboo forests.

Bamboo resource management is a labour-intensive activity. Unless adequate attention is paid, bamboo would become a congested weed among other vegetation. It has to be properly managed for high yields. Field operations include decongestion, saucer weeding, mounding for young clumps, staggered trenches near the clumps to impound moisture, fire control measures, etc. Selective felling of mature bamboo culms duly retaining 6 to 8 culms per clumps to support the clump is a silvicultural operation that ensures health of the clump apart from fetching the desired revenue.

Congestion in Bamboo culms is a common problem, caused by mismanagement. Unrestricted cutting along the periphery of clumps, browsing of young shoots at the edges by cattle, continued removal of young tender shoots for food, digging of rhizomes for making sticks, prevent living rhizomes from spreading outwards. They consequently, develop within the clump, and the new culms so produced create congestion. In very bad cases, clumps appear as tangled mass of twisted and crooked bamboos, impenetrable and unworkable. Proper management is both a preventive and a remedial step.

The other aspect of increasing productivity is linked to the harvesting agency. If the paper mills are engaged as local contractors to extract the bamboo, they do not have any incentive to undertake selective felling. Selective felling requires more time and effort. Moreover, the paper mills are not selective about the maturity of the bamboos as payment is made on the basis of weight. Also bamboos of all age can be used as raw material for the paper mills. Therefore they tend to clear and fell the entire clump. Extraction needs to be supervised by the Forest Department. Ultimately, it is the local forest dependent communities, who as wage labourers would extract the bamboo. So the Forest Department needs to shoulder more responsibility and not absolve themselves of the duty to undertake scientific management and harvesting of the bamboo clumps. This will also increase productivity.

In India, bamboo is not only grown in forests but is also raised in homesteads and farms. However this is of very limited quantity and is essentially restricted to Nagaland and Kerala. It can be planted under agro-forestry system and practices.[8] Despite having abundant bamboo resources, and having the largest area under bamboo in the world, India faces a shortage of bamboo supply for its various industries. This is only likely to widen further, even if just the needs of the paper industry is taken into account and who are the primary users (from industry) of bamboo in India.

Industrialisation of a bamboo-based sector provides potential for development, generating livelihood for the poorest of the poor without any damage to the environment. This opportunity for a win-win situation is yet to be harnessed in India due to a lack of awareness and procedural legal impediments, which can easily be overcome, provided the state plays a pro-active role. The vast and yet untapped potential needs cultivation, primary processing close to the rural poor, integrated processes with one time subsidy towards establishment and transfer of technology and a coordinated, sustained national-level effort towards marketing these products. Bamboo is in the process of being 'rediscovered' in India. Its attributes and potential are increasingly being recognised.

The Planning Commission prepared a report on 'National Mission on Bamboo Technology and Trade Development (NMBTTD)', which is under the Ministry of Agriculture. According to the report, the Indian bamboo industry had a size of Rs. 2,040 crore in 2003, while the domestic market potential was Rs. 4,463 crore, that is, there was scope to increase the size of the industry by at least 2.2 times. Assuming a growth rate of 15-20 per cent, it was projected to be worth Rs. 26,000 crore industries by 2015. The size of the domestic bamboo industry has also been estimated to be Rs. 6,505 crore, according to Pacific Bamboo Resources. The annual bamboo harvest (poles and shoots) in India was slightly over US $23 million according to FAO (Pabuayon and Espanto, 1996).

Constraints facing the Sector and Stakeholder Views

In order to understand the constraints facing the sector, two routes were explored. The first related to looking at secondary documented sources of constraints facing the sector and the second was the various roundtables that

8. Bamboo plantations mainly of *Dendrocalamus strictus* are being raised by different state forest departments. However, there are two main problems: (i) availability of planting stock and (ii) protection of plantation areas.

were conducted in the course of the study in various parts of the country in order to understand the stakeholder responses from the people associated with the resource use, including government officials and representatives from various bamboo-based industries. Several site visits were also undertaken to get a primary view of the process and the practical difficulties encountered. Based on these two routes the findings are briefly mentioned below.

There were several constraints that were identified in the course of the roundtables and field visits as well as from secondary sources. These can be listed as follows:

- The regulatory constraint on transit of bamboo as well as on harvesting from private plantations,
- The irregular supply of bamboo to industries,
- Poor market linkage of the products,
- Technology application for new product design along with testing, certifying of products,
- Lack of an institute on bamboo application and technology,
- Lack of application of known scientific methods in plantation, poor post-harvest treatment, and upgradation of skill formation,
- Waste utilisation, and
- Competition from Chinese products.

The regulatory restrictions on transit and harvest of bamboo are the biggest impediments to the growth of bamboo-based industries and applications. This happens because bamboo is defined to be a tree, when it is a grass and therefore it is treated at par with timber and other forest produces.

The transit pass requirement adds to the delay and increases the cost due to red-tapeism. For example, the cost of one pole of *Dendrocalamus strictus* is Rs. 10 but by the time the pole is available for further processing in Hyderabad city, the price increases to Rs. 40 per pole. This arises from the severe restrictions on movement of bamboo. Failure to get movement licenses forces people to move to other plantation trees like eucalyptus.

Even if the added costs can be factored in, the irregular and delayed supplies hold up production and sometimes it happens indefinitely. Promoting private plantations in bamboo can ease out the problem of irregular supply but until and unless the transit and harvesting restrictions (along with trade restrictions in certain states like Orissa) are tackled the delays and cost escalation will continue to throttle the sector. Bamboos from private plantations also are subjected to these restrictions in most states (West Bengal, Andhra Pradesh and Nagaland being exceptions).

Of the other barriers, in realising the potential of bamboo is the poor market linkage and technology application. New technology and product options need to be developed. Only this can help in encouraging the establishment of manufacturing units. A public interest awareness campaign was felt to be essential for promoting the sector by most of the industry representatives throughout the country. The sector cannot thrive by making handicraft items like baskets anymore. There has to be a movement towards lifestyle products and utility products (not just show pieces). Further therefore, there is the need for market establishment of these products, with product testing for quality being a necessity which will ultimately lead to market acceptability. In this regard, the NMBTTD was mooted by Planning Commission to accord bamboo development a strategic role in rural economy, poverty alleviation and bamboo-based handicrafts and industrial development. However, presently more needs to be done and in a concerted way.

Bamboo technology is not taught in India. So it was felt that there is a need for a bamboo technology institute, which can also impart the desired training. It needs to be located where the bamboo industry shows potential. The institute can also provide immediate solutions to local problems and help disseminate the information after research is carried out for further development. There was the need for a credible certifying organisation that would be accountable on quality issues. This can help the grower who grows appropriate varieties of bamboo that have industrial applications and the institute can spread awareness among states that will promote the use of bamboo. Presently, Central Institute of Plastics Engineering and Technology (CIPET), an autonomous institute under the Department of Chemicals and Petrochemicals, Government of India, is the only agency authorised to do some kind of certification and they certify the tsunami related structures. However, CIPET's core competency is in plastics and not in Bamboo.

Several other constraints also stand in the way of development of this sector in India, like lack of application of known scientific methods in plantation, poor post-harvest treatment, product development and upgradation of skill formation. Low yield per hectare reveals poor management of extant bamboo forests. Inadequate trained manpower and inadequate infrastructure for large scale harvesting in the event of gregarious flowering was also identified as a potential constraint. The bamboo cutters are usually exploited in the present system, especially by the paper mills, with no welfare schemes to benefit them and they work at abysmally low daily wage rates. As a result, many migrate in search of jobs. Bamboo plantation activities over the next five years could generate about 50.4 million man-days of work according to the Planning Commission. In the nursery sector, total estimated employment to be generated every year is to be around 9.7 lakh man-days. Besides this, there will be employment generation in both skilled and unskilled segments in the handicraft sector.

Regulating the Bamboo Sector

The Central laws pertain to forestland which is the property of the government. In other words, the Central laws do not apply to private forests or private plantations. There are three Central acts that govern forest and forest produce. These are the Indian Forest Act (1927), the Forest Conservation Act (1980) and the Scheduled Tribes and Other Traditional Forest Dwellers (Recognition of Forest Rights) Act, 2006. Before proceeding to understand how these laws affect bamboo plantation, harvesting and transportation, it is imperative to outline the objectives that all these three laws purports to achieve.

The Indian Forest Act, 1927 is the single most important piece of legislation on forests. This act has only minor differences with the 1878 Forest Act with the philosophy remaining the same.[9] Two fundamental issues can clearly be identified in the Act. The first pertains to the establishment of absolute state property rights over forests. Towards this end, the classification of forests —into reserved forests, protected forests and village forests—and the legal separation of customary rights as well as the procedure for forest settlement in these was an administrative feature characterising this Act. The second pertained to the control of timber and other forest produce in transit, the duty leviable on them and the collection of drift and stranded timber. The commercial motive and revenue generation remained the guiding principle. Of relevance to this study are the definitional aspects contained in Section 2 and Chapter VII of the Act, which contains detailed and wide encompassing provisions empowering the government (more specifically the state governments) for the control of timber and other forest produce in transit by land or by water. The link between Section 2 and Chapter VII lies in the fact that whatever got defined as a 'forest produce' could be controlled by the state governments through the rules framed by these various state governments in their respective states.

The Forest Conservation Act, 1980, deals with the restriction on allotment of 'forestland' for non-forest purposes and de-reservation of reserved forests. The act is a two-page document, consisting of only five sections. The act clarifies that the term 'non-forest purpose' means the breaking up of or clearing of any forestland for the cultivation of tea, coffee, spices, rubber, palms, oil-bearing plants, horticultural crops or medicinal plants, or any

9. The 1927 Act was promulgated to 'consolidate the law relating to forests, the transit of forest produce and the duty leviable on timber and other forest produce'. A brief genesis of the Act is in Supplementary Note 1.

purpose other than reforestation. The act was a crisis-driven response. The objective of enacting this act was to empower the Central government in directly managing India's forests. The act, is not really a substantive law, it is a delegated legislation, which empowers the Union minister to make the decisions about how to use the forestlands. Further, this act only forbids 'reserve forests' from being de-notified by the states. The need to promulgate this act was felt as remote sensing data of the 1970s showed the adverse consequences of large scale diversion of forestlands to non-forestry purposes (which hit an astounding rate of 1,50,000 hectares per year prior to the 1980s). Again, after forests were transferred to the concurrent list by the 42nd Amendment Act of 1976, the Ministry of Environment and Forests (MoEF) was set up as a nodal Central authority in 1980; the Union government could now directly intervene.

In particular there is no aspect of the Forest Conservation Act, 1980 that is of relevance to this study. However, the Supreme Court case—T.N. Godavarman Thirumulkpad *vs.* Union of India and others (Writ Petition No. 202 of 1995)—was filed in contravention to this act and this case has turned out to be a landmark, one that has perhaps altered the way bamboo ought to be regulated in future. The other Central law is the Scheduled Tribes and Other Traditional Forest Dwellers (Recognition of Forest Rights) Act, 2006, which was notified in the Gazette of India only on January 2, 2007. The rules under this act are not yet notified. The act seeks to undo the historical injustice done to forest dwelling communities and vests property rights on forestland in forest dwelling communities thereby addressing their long-standing insecurity of tenure and access rights. The act gives them access to minor forest produce (including rights to sell), and a stake in the preservation of open spaces. The act provides heritable but non-transferable tenures if they have occupied the lands (up to a maximum of four hectares per family) for three generations from 1930, with December 13, 2005 as the cut-off date. The act gives the right of jurisdiction of *gram sabhas* to settle tribal claims. Of relevance to this study is only the definition of bamboo.

There are two kinds of relevance *vis-à-vis* bamboo that the Central acts deal with. The first is definitional, which defines bamboo as a forest produce by its origin. The second pertains specifically to the harvest and transit rules applicable to bamboo. The issue of trade regulation is contained in some of the state laws. The issue of harvest of bamboo from private lands or private cultivation is also contained in the state laws. Beginning with the second issue

10. According to the Scheduled Tribes and Other Traditional Forest Dwellers (Recognition of Forest Rights) Act, 2006, the forest dwellers have the right to 'collect, use and dispose of minor forest produce' and which includes bamboo, but there is no explicit mention of trade or transit of either timber or any other forest produce. The word 'dispose' is not assumed to mean either trade or transit.

first, that is the harvest and transit in bamboo; it is primarily the provisions of the Indian Forest Act, 1927, that is of relevance.[10] Specifically, by Section 26, removal of any forest produce (harvesting of bamboo) is prohibited in reserved forests, except by the Forest Department (usually in accordance with working plans). In protected forests, (Chapter IV) removal of timber or any other forest produce (including bamboo) is to be done with the written permission of the forest officer or in accordance with the rules framed by the state governments. The rules so framed by the state governments can include among others, granting of licences to persons felling or removing trees or timber or other forest produce from such forests for the purposes of trade, and the examination of forest produce passing out of such forests.

The provisions particularly related to transit are contained in Sections 41 and 42 of Chapters VII.[11] Under clause (1) of Section 41, 'the control of all timber and other forest-produce in transit by land or water is vested in the state government, and it may make rules to regulate the transit of all timber and other forest-produce'. The section is fairly detailed in its ambit, and gives the state governments ample powers including the authority to frame rules for transit, control all river banks (as regards floating timber) and to make rules to prescribe the routes for import, export and other movements of timber or other forest produce from the state or within the state. Thus the state rules regarding prohibition of such movements without pass, issuing of pass and prescribing fees in respect thereof becomes important and almost all states have laws/rules that guide movement of timber and other forest produce.

As mentioned before, the Central laws on forests governed forestlands that were the property of the government. As regards the private forests or private plantations, the state laws, if any, governed the harvesting of bamboo. However, the transit rules applied to bamboo irrespective of its origin as the definition of trees in the Indian Forest Act of 1927 makes bamboo a forest produce. Prior to the harvesting or felling of any tree from private lands, a certificate of origin is required from the state Forest Department, which is issued after due inspection and according to procedures laid out in the state laws that govern private forests for the respective states. Although the procedures vary, the methodology follows this pattern. As seen in the following section, bamboo if included under the definition of a tree, and thus if treated as a forest produce even if its harvesting is from private lands would require a certificate of origin, and which would then form the basis of the issuance of the transit permit. So, Sections 41 and 42 of the Indian Forest Act of 1927,

---

11. Chapter VIII also contains certain related provisions. Chapter VII is titled 'of the control of timber and other forest-produce in transit', while Chapter VIII of the act deals with the 'collection of drift and stranded timber'. There was no particular need to have two different chapters on very similar aspects, and the former chapter was quite sufficient.

and thereby all rules framed on transit of forest produce by the states, will be applicable to bamboo as long as it is a forest produce.

Conclusion

Despite having fairly detailed estimates on the size and potential of the various market segments of the bamboo-based industries or where bamboo is used, and despite they being estimates of a Planning Commission document, the appropriateness and reliability of the market potential, along with its availability and method of estimation is circumspect. As a result these estimates, duly quoted, are nevertheless questioned here. The process of consultation with the stakeholders—people associated with the resource use, including government officials and representatives from various bamboo-based industries, along with the site visits to get a primary view of the process and the practical difficulties encountered helped us formulate the difficulties encountered.

The most important bottleneck was identified as the regulatory bottleneck, and as long as this was not meaningfully addressed, the sector cannot grow beyond a certain threshold level. In general, there seemed to be a strong promotional role of a governmental organisation, which would help generate awareness on bamboo products, run a nationalised campaign and help develop product-market linkages, apart from handholding the sector in the initial stages. The government initiatives and agencies created for the development of the sector was described and although there seems to be overlapping of jurisdictions, the sector is presently at such a nascent stage of development that there cannot be shortages of initiatives. What perhaps is lacking is a more concerted effort and better planning, and we are of the opinion that either of the missions, NMBTTD or NMBA (National Mission on Bamboo Applications), and perhaps the latter, should be converted into a permanent 'board' for the development of the bamboo sector in India, in lines of the 'coffee board' or the 'tea board'. This is especially called for because the report argues for bamboo cultivated in private plantations and it is hoped that in future bamboo will be treated at par with other plantation crops once the viability of private plantations is demonstrated in different parts of the country.

References

Albion, R.G. (1926). *Forests and Sea Power*. Cambridge: Harvard University Press, Massachussets.

Agarbatti Stick Production Under Andhra Pradesh Community Forest Management Project, INBAR-CIBART Documentation Centre, Bangalore, May 2006.

ASSOCHAM (2007). "Replace Wood Products With Bamboo's to Help Govt. Save Rs. 7000 Cr. P.A", *Mimeo*, October 28.

Baden-Powell, B.H., (1882). *A Manual of Jurisprudence for Forest Officers*. Calcutta: Government Press.

Balaji, S. (1991). "Agro-forestry for Prosperity", *Forest News* (1) (3): 9-11. Tamil Nadu: Forest Department.

Banik, R.L. (1986). "Macro-propagation of Bamboos by Pre-rooted and Pre-rhizomed Branch Cutting", *Bano Bigyan Patrika* 13(1/2): 67-73.

Basu, S.K. (1985). "The Present Status of Rattan Palms in India—An Overview", in K.M. Wong and N. Manokaran (eds.), *Proceeding of the Rattan Seminar* 77-90, FRI, Kepong, Malaysia.

Bhatt, K.M. (1992). "Changing Scenario of Rattan Trade in India", in S. Chand Basha and K.M. Bhatt (eds.), *Rattan Management and Utilisation*, Proceedings of the seminar held on 29-31, January, in Trichur, Kerala, pp. 335-339.

Cane and Bamboo News, Quality bulletin of CBTC 1(4).

Champion H.G. and S.K. Seth (1968). *Revised Survey of the Forest Types of India 1-402*. Delhi: Government of India Press.

Cleuren, H.M., and A. B. Henkemans, (2003). "Development of the Bamboo Sector in Ecuador: Harnessing the Potential of Guadua Angustifolia", *Journal of Bamboo and Rattan* 2(2), September.

CRISIL (2006). "Demand Fundamentals Support Asset Mortgage Quality", *Crisil Insight*, January.

Customs General Administration of China (2004). *China Customs Yearbook*. Beijing: Customs General Administration of China Publishing.

FAO (1997). "Asia-Pacific Forestry Sector Outlook Study: Country Report—Malaysia", *Working Paper* No: APFSOS/WP/07. Forestry Department Headquarters, Peninsular Malaysia, Kuala Lumpur, Malaysia and Forestry Policy and Planning Division, Rome. Bangkok, FAO Regional Office for Asia and the Pacific.

FAO (2006). "Global Forest Resources Assessment 2005: India Country Report on Bamboo Resources", *Working Paper* No. 118, Forestry Department, FAO, Rome.

Forest Survey of India (1999). "State of Forest Report 1999", Dehradun: Ministry of Environment and Forests.

Gadgil, M. and R. Guha (1992). *This Fissured Land: An Ecological History of India*. New Delhi: Oxford University Press.

Ganapathy, P. M. (1997). "Sources of Non Wood Fibre for Paper, Board and Panels Production: Status, Trends and Prospects for India", *Working Paper* No: APFSOS/WP/10, Asia-Pacific Forestry Sector Outlook Study Working Paper Series, Asia-Pacific Forestry Commission, Rome.

Gangopadhyay, P. B. (2003). "Bamboo Resources as a Rural Livelihood Option in Madhya Pradesh, India", Paper submitted to XII World Forestry Congress, Quebec City, March.

Guha, R. (1983). "Forestry in British and Post-British India: An Historical Analysis", *Economic and Political Weekly* 29 October and 5 November, pp. 1882 and 1940.

———. (1990). "An Early Environmental Debate: The Making of the 1878 Act", IESHR 27: 65.

———. (1991). *The Unquiet Woods: Ecological and Peasant Resistance in the Himalayas*. New Delhi: Oxford University Press.

ICFRE. (1998). *Timber/Bamboo Trade Bulletin*. March, No. 14. Dehradun: Directorate of Statistics, ICFRE. pp. 31.

Kumar, Arun and Cherla B. Sastry (1999). "The International Network for Bamboo and Rattan", *Unasylva* 198 on *Non-wood Forest Products and Income Generation*, FAO.

Lakshmana, A.C., (1994). *Rattans of South India*. Bangalore: Evergreen Publishers.

Madhab, Jayanta, (2003). "The Green Gold: Under Exploited Wealth of the North East India", *Dialogue*. 5(2), October-December.

Maogong, Zhong and Liu Chang (1999). "Retrospects and Prospects on Development of Bamboo Sector in China", in *Forestry Economy* 3: 51-62.

Mohd. Ali, Abdul Razak and Abd. Latif Mohmod (1998). "Non-wood Forest Resources and Products: Management and Research Considerations", in M.N.B. Nair, Mohd. Hamami Sahri and Zaidon Ashaari (eds.), *Sustainable Management of Non-wood Forest Products*. Proceedings of an International Workshop held at Universiti Putra Malaysia, Serdang, 14-17 October 1997. Serdang: Universiti Putra Malaysia Press.

Naithani, H.B. (1993). "Contributions to the Taxonomic Studies of Indian Bamboos", PhD. Thesis, Vol. I. Srinagar, Garhwal: H.N.B. Garhwal University.

National Mission on Bamboo Applications, New Delhi, various documents.

Negi, S.S. (1996). *Bamboos and Canes*. Dehradun: Bishen Singh Mahendra Pal Singh.

Negi, S.S. and H.B. Naithani (1994). *Hand Book of Indian Bamboos*. Dehradun: Oriental Enterprises.

Pabuayon, I.M. and L.H. Espanto (1996). "The Philippine Rattan Sector: A Case Study of an Extensive Production System", Bamboo and Rattan Seminar/Workshop, 28 June.

Planning Commission (2003). "National Mission on Bamboo Technology and Trade Development". Delhi: Government of India.

Press Information Bureau, (2004). "The Wonderful Natural Resource", Press Release, March 25.

Punhani, R.K. and K.S. Pruthi (1991). "Substitution of Wood in Building Some Alternative Forest based Materials and their Technology", National Symposium on *Substitution of Wood Building* (SWOB), Roorkee.

Savur, Manorama (2003). *And the Bamboo Flowers in the Indian Forests: What did the Pulp and the Paper Industry Do?* Vol. I and II, Delhi: Manohar Publications.

Seshadri, P. (1985). "Intercropping of Bamboo (Dendrocalamus strictus Nees) with Soybean (Glycine max L.) Heril: An Agroforestry Study", PhD. Thesis. Coimbatore: Tamil Nadu Agricultural University.

Shanmughavel, P., K. Francis and M. George (1997). *Plantation Bamboo*. Dehradun: International Book Distributors.

Sharma, S.N. (1988). "Seasoning Behaviour and Related Properties of Some Indian Species of Bamboo", *Indian Forester* 114(10): 613-621.

Singhal, R.M. and P.B. Gangopadhyay (1999). *Bamboo and Its Database in India*. Dehradun: ICFRE Publications.

Saxena, N.C. (1997). *The Saga of Participatory Management in India*. Jakarta: CIFOR Special Publication.

Singh, G. (1995). *Environmental Law*. New Delhi: Lawman (India) Pvt. Limited.

Singh, K. (1994). *Managing Common Pool Resources*. New Delhi: Oxford University Press.

Singh, C. (1986). *Common Property and Common Poverty: India's Forest Dwellers and the Law*. New Delhi: Oxford University Press.

Stebbing, E.P. (1922-27). *The Forests of India*, Vols. I - IV, London: John Lane.

Subramaniam, K.N. (1998). "Bamboo Genetic Resources in India", in K. Vivekanandan, A.N. Rao and V. Ramanatha Rao (eds.), *Bamboo and Rattan Genetic Resources in Asian Countries*, IPGRI-APO, Serdang, Malaysia.

Tewari, D.N. (1992). *A Monograph on Bamboo*. pp. 1-498. Dehradun: International Book Distributors.

Wagh, R. and J.C. Rajput (1991). "Comparative Performance of Bamboo with the Horticultural Crops in Konkan", in *Bamboo in Asia and Pacific*. Proceedings of the Fourth International. Bamboo Workshop, 27-30 Nov. 1991. Chiangmai, Thailand, FORSPA Publication-6. Canada. IDRC and FORSPA, Bangkok, Thailand, 1994 : 85-86.

13

Agroforestry for Minimising Food, Fodder and Fuel Scarcity

S.K. KARN and SARITA MUTHA JAIN

Introduction

Low and erratic rainfall and intense solar radiation coupled with high wind velocity is the characteristic feature of Indian arid zone. Recurring drought and famines are common features in the region. In view of the natural calamities in the region, farmers have adopted traditional agroforestry since ages. Different traditional agroforestry systems are promoted in the region to combat these vagaries (Malhotra, 1984). Tree crop combination or agriculture and forestry technologies create more integrated, diverse, productive, profitable, healthy, and sustainable land-use systems (Anderson and Sinclair, 1993; Prinsley, 1992). The farmers of this region maintain and promote the trees like *P. cineraria, Acacia nilotica, Ailanthus excelsa* and *Ziziphus nummularia* on their farmland, which help them to sustain during the droughts. Due to limited resources in arid and semi-arid regions, benefits from the agroforestry system are largely depending on efficient and judicious management of soil and water resources (Parton *et al.*, 1992). Selection of appropriate combination of trees and crop species and practices such as pruning/lopping and spacing regimes are the important aspects of agroforestry. The trees raised in the agroforestry have been planted keeping in view the socioeconomic and ecological factors. With the increasing demand of fodder and fuel wood, agro- forestry has gained priority in the region. Association of trees with agricultural crops affects crop productivity (Aggarwal and Kumar, 1990; Brenan and Kessler, 1995) through changes in soil properties (Kater *et al.*, 1992) and water conservation (Gupta and Saxena, 1978). It ultimately provides many tangible and intangible benefits to the farmer.

Materials and Methods

Site conditions

Desert soils, which are characterised as sand, poor in humus and water holding capacity and high salt content commonly xerophytic species are found. e.g., *Prosopis cineraria (khejri), P. chilensis, Acacia tortilis* (Israeli *babul*), *A. senegal (kumat), A. nilotica (babul), Ziziphus mauritiana (ber), Salvadora* spp. (*pilu*).

The study was conducted in the western parts of Rajasathan in the parts of Jodhpur, Pali, Nagaur and Udaipur. The mean annual rainfall of the site is 350 mm. The maximum temperature rises as high as 48^{0}C in the summer and the minimum drops to 0^{0}C in the winter. Average wind velocity in the summer months is 20-30 km^{-1}. Study sites were the farms of the local villagers. The soil condition of the farm was flat land with loamy sand soil (coarse loamy, mixed hyperthermic family of typic camborthids according to US soil taxonomy) underlain with a thick concretion of calcium carbonate at a depth of 75 cm (in the Jodhpur and Pali districts). The soil pH is 7.8. Soil moisture storage in the upper 75 cm layer varies from 120 mm at 0.01 MPa to 35 mm at 1.5 MPa.

Climatic conditions in arid regions, annual rainfall is approximately in the range of 100 mm to 400 mm with an interannual variation of 50-100 per cent.

In the Jaisalmer region *Ziziphus nummularia* (*bordi, jharber*) is grown with agricultural crops. *Prosopis cineraria* (*khejri*) trees are also found in some fields but are very few in number. In Barmer, crops are grown in association with *khejri* and *bordi*. In the Jodhpur and Nagaur belt crops are mainly grown in association with *khejri*; *bordi* plants are very few in number. In the Jodhpur-Pali-Udaipur belt the dominating tree species are *Acacia nilotica* var. *cupressiformis is* commonly called as *khajuria babul*. Both the *kharif* and *rabi* crops are taken with these trees. In the Jaipur (*ardu*) Dausa belt the *Ailanthus excelsa* trees are planted for fodder. These trees are generally planted on the boundary of field as boundary plantations.

Prosopis Cineraria-based Agroforestry Models

P.cineraria commonly called as *khejri*, is the most important tree to combine with agricultural crops in the rainfall zone of 250-400 mm. Its density in agricultural fields varies with the rainfall and site conditions. Optimum density of 80-120 trees per ha is observed in districts of Churu, Sikar, Jhunjhunu and Nagaur. Its suitability in agroforestry system stands from the fact that a well-lopped *khejri* tree recovers its full crown in a period of nine months. The crown develops by the end of September. Thus when the *kharif* crops mature the foliate is lopped to facilitate better growth of *kharif* crop. In case, the fields are to be used for *rabi* crops like wheat and mustard, lopping is continued slowly during summers and fast during monsoon.

Khejri's diversity makes it a valuable "companion" to agricultural crops. *Khejri* is a nitrogen fixer, which means it improves soil quality by making nitrogen in the soil more available to other plants. Its leaves further improve the soil by adding organic matter. With a taproot that can extend more than 100 feet deep and an extensive root mass, *khejri* helps to stabilise the sandy desert soil and shifting sand dunes. It can serve as a windbreak, protecting farms from strong desert winds, and its wood is excellent for firewood and charcoal.

Khejri is a symbol of sustainable socioeconomic development for the arid Indian deserts. *Khejri* has a very deep taproot system, hence it does not generally compete with the associated crops. The improved physical soil conditions compared with higher availability of nutrients under the canopy explain the better growth of the crops associated with it. Due to its extensive root system it stabilises shifting sand dunes and is also useful as windbreak shelterbelt and in afforestation of dry areas. The trees not only boost the growth and productivity of companion plants, but also provide fuel, fodder, food, small timber, medicines, gum and tannin. Its foliage is a nutritive fodder for animals and the wood is of good quality for domestic fuel purposes. Unlopped trees produce green, immature pods (*sangri*) used as a vegetable (fresh and also dried), and ripe pods (*khokha*) are used for fresh consumption and for the preparation of flour. Khokha flour is mixed with wheat flour to make bread (*chapati*) and bakery products. Foliage and pods of *khejri* are very nutritious, being rich in protein, carbohydrates and mineral matter. Green leaves of *P. cineraria* contain 14-18 per cent crude protein, 13-22 per cent crude fibre and about 6 per cent ash, with a high calorific value of 5000 Kcal. Unripe pods are also nutritious and are used to prepare curries and pickles. *Khejri* fruits or pods are locally called *sangar* or *sangri*. The dried pods locally called *khokha* are eaten. Dried pods also form rich animal feed, which is liked by all livestock. Green pods also form rich animal feed, which is liked by drying the young boiled pods. The dried green *sangri* is used as a delicious dried vegetable which is very expensive (Nearly Rs.150 per kg in market). Locals use the *sangri* in preparation of curries and pickles.

Generally the tree is lopped heavily during winter season. In double cropped area, in the month of October, whereas in single cropped area the lopping is done from November to December. A mature tree about 7 m height and 83 cm (dbh; diameter breast height) when fully lopped yielded around 59 kg leaf fodder whereas 66 and 33 per cent lopping could yield 28.5 and 20 kg per tree, respectively. Lopping at an interval of four years produced 238 per cent higher leaf fodder than the annual lopping. It is reported to yield maximum leaf fodder production when trees are lopped in winter season. The nutritive value of this leaf fodder is also of very high order. Total returns from *khejri* over sole agricultural crop vary between Rs. 6,400-22,400.

Social Importance

During Vedic times, *khejri* wood was used to kindle the sacred fire for performing a *yajana*. The Hindu epics, the Ramayana and the Mahabharata, mention the usefulness and significance of this tree. Married men and women worship *khejri* (*jand*) tree, in an elaborate way. The Bishnois, a community in Rajasthan, would not cut khejri trees even in their agricultural fields.

Ailanthus Excelsa (*Ardu*)-based Agroforestry Models

It is an important lightwood agroforestry species grown around agriculture fields. Trees are planted on field boundary at a spacing of 3 m or 5 m. Wider spacing is more popular as trees have sufficient space to produce more leaf biomass. Fodder crops or agriculture crops, as per requirement of the farmer, are intercultivated, leaves are highly palatable and provide protein rich nutritious fodder (crude protein 12.82 per cent, Ca 2.11 per cent, P 0.24 per cent, on dry matter basis) for sheep and goats, and is said to augment milk production of cattle. Average yield of leaf fodder at 4, 8 and 12-13 years age is 20 kg, 85 kg and 200 kg per tree per lopping. Lopping for foliage is done twice a year during the main cropping season. A single tree, on an average produces 500 to 700 kg of fodder twice a year under irrigated situation. *Ailanthus excelsa* has an excellent market in the Jaipur city where its leaves are sold as hot cakes. Green leaves are also sold in the market, especially for nourishing stall fed goats. Farm forestry of *A. excelsa* is more popular amongst farmers. Farmers are getting more yields of crops associated with this species. They are also generating extra revenue from *A. excelsa*. It is reported that crops grown in combination with this species suffer no adverse effects. The agriculture crops like maize, pulses, groundnut, wheat, millet, barley and mustard are intercultivated with this species. The average, additional income received from planting *Ailanthus excelsa* is Rs. 7,600/ha/year. The rotation of *A. excelsa* is considered as 20 years but it was found to be 30 years or more.

It provides revenue to the owner and employment to other people engaged in its marketing. In Jaipur low quality cricket bats, stumps and stools are marketed. One cubic metre wood is available to them at Rs. 1,750, which can give 250-350 normal-sized bats. Small toys such as elephants and camels are sold at Rs. 12/dz (dozen) and Rs. 50/dz. The difference in the rates is due to the labour cost. A carpenter makes 500-600 pieces of elephant in a day while only 150 pieces of camels can be made in the same time. These toys are sold at the market price of Rs. 3-4 and Rs. 6-7 respectively.

Ziziphus (*Ber*)-based Agroforestry Models

Ziziphus rotundifolia (*ber*) is an important fruit tree of arid area. *Z. nummularia* (*jharber*) that grows naturally at the farmers' fields and does not compete with the crops. However, budded *Z. mauritiana* (*Ber*) has also been found to grow very well under arid conditions when planted either with water harvesting system or with supple-mental irrigation. Three-year old plantation of *ber* at a density of 400 plants ha^{-1} in association with *moong* beans performs very well despite of seasonal rainfall of less than 220 mm only.

The inventory of this system shows that it can provide, round the year supply of fodder for five goats/sheep and fuel wood for a family of four members besides efficient nutrient cycling and increase in economic stability. *Ber* fruit and dry leaves have great importance during drought years.

This system provides nearly 2.5 t/h fodder as *ber* leaves and *moong* crop residue. *Ber* leaves are considered to be the best fodder for goat and sheep in this area. Also *moong* crop straw can be helpful to reduce the protein deficiency, which is common in this area due to the dominance of grassy fodder. Moreover, this system supplies fodder round the year, i.e., *moong* crop residue from October onward and *ber* leaves from February onward. The quantity of fodder available from one hectare is sufficient for five goats/sheep for one year. The goats/sheep fed on this fodder can provide about one tonne of excreta. In this hot arid region, applying nutrient through organic sources have advantages over inorganic forms because of the lesser nutrient loss in this hot and arid climate. Organic sources increase organic matter and water holding capacity of the soil. Farmers can neither think nor are capable to apply inorganic fertilisers in this highly risky crop producing area. So such practice is needed for building up of soil fertility.

Pruning of *ber* trees is essential for maintaining growth and fruit production during subsequent years. Besides, it makes the fuel wood available from *ber* tree without disturbing the eco-balance. Having calorific value of 4878 Kcal/kg wood and considering 20 per cent heat conversion efficiency and 3.0 kg/day fuel wood requirement, the wood obtained by pruning is sufficient for cooking the food, round the year for a family of four members.

Socioeconomic Impact

This system increases the stability and income of marginal farmers by providing round the year work to the farmer who, otherwise remain idle for eight months after harvesting the crops in sole cropping system. The farmers also earn during drought years (60 per cent of the normal year in terms of fruit, fodder and fuel wood yield). The prevailing rates of *ber* range from Rs. 20 per kg and *moong* bean Rs. 35/kg.

Acacia Nilotica (*Babul*)-based Agroforestry Model

Acacia nilotica has been grown since ages in semi-arid and arid regions. This species is grown under agri-silviculture and silvo-pastoral systems. It is prevalent in 450-500 mm rainfall zone in the districts of Ajmer, Jaipur, Dausa and Pali. This is largely under irrigated conditions where both *kharif* and *rabi* crops are taken. There are several types of crop combinations and crop sequences prevalent depending upon the availability of irrigation water. Generally in this area *Acacia niltotica* var. *cupressiformis* (*khajuria babul*) is found randomly planted in the farmer's fields. Trees are planted along borders. They are lopped periodically for fuel wood, and their leaves and pods are also harvested for fodder. In addition, normal littering adds as organic matter to be used by the associated agriculture crops. Row plantation along with agriculture crop/grass is also common in several areas. In another practice, scattered plants growing naturally are retained and maintained by farmers. Wind erosion is the major problem faced in drier parts as it adversely affects agriculture and grazing resources. Species have been grown along roads, around reservoirs and around large farms. It is a common species in semi-arid and arid regions combined in a multilayered cropping system of home gardens. *Acacia nilotica* regenerates naturally on farmlands from seeds disseminated by birds, animals and wind. It is a traditional practice followed by farmers to retain and maintain a few plants in the field. The number may vary from 20 to 25 plants per hectare. This practice is most common under arid situations. Plants and pods are lopped regularly for fodder requirement. The bark and pods which contain tannin are widely used in the leather industry. Their tannin content varies from 12 to 20 per cent. Leaves and pods contain 15 per cent crude protein which is a good source of nourishment to animals. The tree is a good source of gum, a valuable commodity used in home consumption, in the manufacture of matches, inks, paints and confectionery. Gum production generates additional income to the farmers. Total returns from *Acacia nilotica* over sole agricultural crop vary between Rs. 6,490 and Rs. 7,880/ha.

Conclusion

Low and erratic rainfall in arid and semi-arid regions of western India has led to repeated droughts in the arid parts of the country. It is the major problem for sustainable production for the farmers of the region. Diversification into products from trees may provide some insurance against such uncertainties. The integration of trees into farming systems 'agroforestry' has the potential to improve both sustainability and profitability in substantial quantities to reduce the pressure from our forest resources for fuel and fodder. Tree planting on farm helps to protect the environment and maintain productivity. Other than supplying products such as food, fodder and timber, it also helps in salinity control, erosion control, conservation of biodiversity, improving soil fertility, etc. The integration of trees into the farming system can potentially provide income to the family. Natural calamities in world prices for commodities, droughts and other climatic events result in unstable farm income. A tree can support both climatic and market instabilities. A small landholding owner cannot produce a economically viable large volume for high trade, but is sufficient for fulfilling his own family and farm needs. Doors and windows, agricultural implements, small toys (marketed on small scale in the region) help him. Another important contribution of trees under agroforestry is fodder, animal husbandry, a simultaneous occupation, helps the livestock to survive in the scarcity periods, which in turn supplies milk and milk products for extra income. Trees in the arid areas are well adapted to the xeric conditions. The main contributions briefly are: increased and more stable food production directly due to introduction of trees (fodder, fruits) or indirectly due to increased soil fertility and the sustainability of agriculture; more products and services; higher income due to sale of tree products; reduced dependence on external sources for key agricultural inputs (fertilisers) and subsistence products (fuel wood, building materials); reduced pressure on grazing through intensification of fodder and animal production; enhanced microclimates; reduced pressure on remaining forests; and improved soil structure and fertility, with important effects on crop yields.

References

Aggarwal, R.K. and Praveen Kumar (1990). "Nitrogen Response to Pearl Millet Grown on Soil Underneath *P. Cineraria* and Adjacent open site in an Arid Environment", *Annals of Arid Zone* 29(4): 289-293.

Anderson, L.S. and F.L. Sinclair (1993). "Ecological Interaction in Agro Forestry Systems", *Agroforestry Abstracts* 6: 57-91.

Brenan, H. and J.J. Kessler (1995). *Woody Plants in Agro Forestry Systems of Semi-arid Regions with an Emphasis on the Sahelian Countries*. Berlin, Springer-Verlag. p. 340.

Gupta, J.P. and S.K. Saxena (1978). "Studies on the Monitoring of Dynamics of Moisture in the Soil and the Performance of Ground Flora under Desertic Communities of Tree", *Indian Journal of Ecology* 5: 30-36.

Kater, L.J.M., S. Kante and A. Budelman (1992). "Karite (*Vitellaria paradoxa*) and nere (*Parkia globosa*) associated with crops in South Mali", *Agroforestry Systems* 18: 89-105.

Malhotra, S.P. (1984). "Traditional Agroforestry Practices in Arid Zone of Rajasthan", in K.A. Shankarnarayan (ed.), *Agroforestry in Arid and Semi-arid Zones*. Jodhpur: CAZRI. pp. 263-266.

Parton, W.J., D.S. Schirul, C.W. Cole and D.S. Ozima (1987). "Analysis of Factors Controlling Soil Organic Matter Levels in Great Plain Grasslands", *Soil Science Society of America Journal* 1173-1179.

Parton *et al.* (1992). "Relationships between the Farm Household and Adoption of Improved Technologies in Semi Arid Tropics of Eastern Kenya", *Journal of Farming Systems Research and Extension*.

Prinsley, R.T. (1992). "The Role of Tree in Sustainable Agriculture: An Overview", *Agroforestry Systems* 20: 87-115.

IV

Regional Issues in Forest Management

14

Intra-Community Conflict Resolution Mechanism and the Role of Social Capital

Lessons From a Joint Forest Planning and Management Project in Karnataka

OZMOND ROSHAN D'SOUZA

Introduction

This paper is a part of a larger study of joint forest planning and management (JFPM) in Karnataka. The field study was based in Shimoga forest circle which includes the whole of Shimoga revenue district and parts of Chikmagalur and Davanagere revenue districts. The main focus of the study was to understand the influence of social capital on JFPM and vice-versa. The concept of social capital which in the broad sense refers to features of social organisations, such as networks, norms and social trust that facilitate, co-ordinate and cooperate for mutual benefit (Putnam *et al.*, 1993), was applied to study the programme. This concept of social capital is helpful in understanding the institutional performance of community-driven/based development projects. The understanding of the theory of social capital becomes all the more important in studies related to common property resources, because social capital holds the key to understanding multi-stakeholders' relationship/networks and its influence on institutional performance. Social capital is a multifaceted concept and can help explain different aspects of networks in a community-driven/based development project that involves four major stakeholders in the form of the community/ies (beneficiaries), funding agencies (benefactors), implementing and facilitating agencies. While bonding social capital explains networks within one stakeholder group, generally, the community/ies, bridging and linking social capital explain the horizontal and vertical networks between and among the different stakeholders.

The present study seeks to understand, through the application of social capital, conflict resolution mechanism, in a particular situation, in which complications arise over benefit sharing among four different geographically divided smaller communities, which operate as a 'single revenue entity' under the JFPM Programme. JFM/JFPM was a result of a feeling that forest management programme needs to be reoriented in such a manner that they respond to the needs of the local community and that the forest produce like fuel wood, minor forest produce, timber, etc., are made available to the village community preferentially, in addition to a share in the proceeds.[1]

Village Profile and the Brief History of the Village

This particular village, Kanagodu-Nadamanchale, was selected for the study as it fulfilled the stipulated selection criteria. Firstly, the village forest council (VFC) in this village was established in 1996 under Overseas Development Administration (ODA), UK funded Western Ghats Development Programme (WGDP) and continued under National Afforestation Programme (NAP), thus fulfilling the first criterion of being established at least five years before 2005 and continued under another project. Secondly, this village was selected to represent Sagar forest division (Madhya Pradesh), to fulfill the criterion of representing one of the administrative divisions of Shimoga forest circle. There are three forest divisions in Shimoga forest circle. Each of the two villages/ VFCs selected for case studies belong to other two forest divisions. Thirdly, the number of households was in the range of 100+, ideally suited for carrying out household survey through census method. Lastly, this village had witnessed incidents of forest fire in the past, fulfilling the last criterion of any incidents requiring collective action to assess the different forms and levels of participation.

Kanagodu-Nadamanchale is a revenue village consisting of five hamlets—Kanagodu, Purlamakki, Station-Balegundi, Doddabylu and Odakere. Station-Balegundi has no forest fringe or plantation but was included under the VFC just because it was a part of the VFC revenue village. Kanagodu in Kannada language means land of the forests. This village belongs to Ulluru *gram panchayat* and *Kasaba Hobli* and Sagar *taluk* of Shimoga revenue district. The village is adjacent to National Highway 264 on the way to Sagar from Shimoga. This village comes under the jurisdiction of Sagar forest range and division of Shimoga forest circle. The *Kanagodu Grama Aranya Samithi* (VFC) was established on 11th November 1996 under the ODA funded, WGDP and is continued now under NAP and monitored by Forest Development Agency (FDA) with Reg. No. 11:96-97. The size of the forest plantation is 138.5 hectares spread over survey nos. 10, 18, 19, 22, 37, 38, 45 and 45.5F in a range of 5 Km. Kanagodu is around 8 km from Sagar and around 67 Km from

1. As stated in the Preamble of the Karnataka Joint Forest Planning and Management Resolution, 1993.

Shimoga district headquarters. Purlamakki is about a kilometre or so from Kanagodu. Odakere is adjacent to Purlamakki. Doddabylu is around 4.5 km and Station-Balegundi is around 5.5 km. A railway track runs through the borders of all the five hamlets. This track between Sagar and Shimoga is under conversion from metre-gauge to broad-gauge.

The revenue village is spread over 1359.2 acres of which 577.57 (385.45 acres in community land) acres is revenue land, 45.1 acres of cultivable land and 326.13 acres of forestland. There are 4 irrigational tanks in the village, 6 open wells and 9 borewells. There are 2 lower primary schools—one in Kanagodu and the other in Station-Balegundi. The revenue village has 5 temples of which 2 are popular in and around the district. There are 25 astra-ole and 19 gobargas plants, most of them in Kanagodu hamlet. The major caste groups in the village are Brahmins, Madiwalas and Madigas.

In the 1980s, there was a movement in Sagar *taluk* and the surroundings, against the Mysore Paper Mills Ltd. (MPM), a Karnataka government enterprise. The MPM would take vacant lands from the Karnataka forest department and raise its own plantation of *acacia* and *nilgiri* to supply the logs to be made into pulp and then finally paper. The people of Malnad region and the Karnataka Forest Development Corporation (KFDC) did not approve of the activities of the MPM, who acted with purely profit motive. They were accused by the locals of destroying the ecological balance by deforestation and raising of monoculture plantation, mainly acacia and nilgiri, both of which would harm the soil and water resources. The MPM was supposed to give 12.5 per cent from the harvested revenue to the local *gram panchayat*. Due to the ignorance of the villagers and no proper agreement, the MPM refused to part with any money.

Though the villagers in Kanagodu knew of the ill effects of the monoculture plantations, in 1985 they decided to invite the MPM officials to start a plantation. This was done mainly to ward off encroachers coming from distant places displaced by different development/hydel projects who found vast vacant lands surrounding the village very attractive. The encroachers made attempts to woo the villagers and allow them to settle down. Many encroachers even pressurised them through politicians. The villagers were of the view that the only way of keeping the encroachers at bay was by inviting the MPM and KFDC to raise the plantations. They were initially successful in keeping the encroachers out too. The villagers were pleased to see the trees grow and greenery all around, but after a few years, the MPM authorities in early 1994 harvested the whole plantation and left the place. These plots almost resembled a desert. That is when the villagers started fearing the potential problem of encroachers and not knowing what to do with the vacant land.

One of the villagers, now the VFC president, chanced upon a piece of news in a local daily about the ODA-WGDP. The news related to the inauguration of a VFC in the neighbouring Karwar district by the forest minister. Interestingly, the name of that village was *Kanagodu*! He went to the Divisional Forest Office to explore the possibility of starting a VFC in his own village. The initial response from the Forest Department was not encouraging, but upon persuasion, a few interested individuals were successful in getting the nod of the forest officials to start a VFC. Slowly the process of the formation of the VFC started in the village.

For the first five years of the project, *Shanthinikethan*, an NGO was actively involved with the VFC. Kanagodu VFC has not had elections to the executive/managing committee since inception. The president and the members were unanimously elected during the micro-planning process. Presently another committee with the same president is continuing with slight modification in the membership as per the requirements under the FDA guidelines.

History

Kanagodu, meaning forestland, got its name from the lush green forest surrounding it. People have been residing in this region since ages. Historical material found in and around the village suggests that this settlement is at least 400 years old, dating back to the days of legendary King of Keladi dynasty, Shivappa Nayaka. Way back in 1918, the first recorded event was written when B.R. Seetharam, well-versed in medicinal herbs, would treat sick domestic animals. It was recorded in 1925 that the aggregate rainfall that year was around 20 inches. There was no dearth of forest resources and people would use these for furniture making and the economic activities revolved mainly around forest resources.

During the 1930s four Brahmin (*Havyaka*) families came and settled in this village, what is now known as Kanagodu, mostly depending on agriculture. These four families have now multiplied into around 20 families and there are also some other families who have joined the village over the years as labourers and artisans. The village boasts of a famous local freedom fighter M. Ganapathi Rao. This village also has an ancient temple named after Shri Neelakanteshwara which is now renovated.

Not much is known about the other four hamlets. Settlements came around the railway track (between Shimoga town and Talaguppa) that was laid in the early 1920s. Odakere hamlet had around 8 families residing, but 7 families left the hamlet in early 1990s for another village to pursue agricultural activities. Purlamakki, was an extension of Brahmana Manchale, a neighbouring hamlet, with only a few Brahmin families. People from other castes and communities joined the hamlet in the late 1960s and the 70s.

Doddabylu, meaning vast plains, has no particular history. Villagers say that a few Madiwala families came and settled in this hamlet, a vast plain in the midst of lush green forests to pursue agricultural and other allied activities in the late 19th and early 20th centuries. Over the years a few other families have joined the hamlet. Station-Balegundi is an extension of Balegundi hamlet, after the village was bifurcated and added to Nadamanchale revenue village, a prefix station was added, to differentiate from the actual Balegundi hamlet. There is a railway station in this part of the hamlet and hence the prefix. Many families work for the railways. A couple of families have been working for two generations as gang men in the railways.

Significance of the Study

There are numerous studies on conflicts and conflict resolution mechanisms related to common property resources. The significance of the present study lies in the fact that it attempts to capture the processes of conservation and development activities carried out by a 'community' made up of four smaller communities divided geographically but united administratively. The conflict and conflict resolution mechanism recorded in the study are not over the common property resources itself, but over the sharing of benefits that the funding and implementing agencies provide for the good work done. While two smaller communities (Kanagodu and Doddabylu) are closer to the conservation sites, the other two (Purlamakki and Station-Balegundi) are not. The study attempts to understand the conflict regarding the sharing of benefits under the development component of the programme and conflict resolution mechanism that follows.

The study further highlights the nature of the conflict, i.e., communities with lesser role in conservation activities demanding greater share in the benefit package. Furthermore, the conflict resolution mechanism is understood and explained from the social capital perspective.

Research Methodology

The present research is a mixture of quantitative and qualitative analysis and a descriptive study. Social Capital Assessment Tool (SCAT), a field-tested set of indicators and methodologies that measure levels of cognitive and structural social capital in communities designated as beneficiaries of development projects developed by Krishna and Shrader (1999)[2] was used to operationalise the research. The SCAT draws on qualitative and quantitative data useful for determining baseline levels of social capital and monitoring progress over the course of project implementation, thereby permitting the

2. Mohan and Mohan (2002).

analysis of the relationships between development indicators and social capital accumulation.[3] The SCAT consists of: (1) a community profile which integrates participatory qualitative methods with community survey instruments to assess various dimensions of community-level social capital, including community assets identification, collective action, solidarity, conflict resolution, community governance and decision-making, institutional networks, and organisational density; (2) a household survey which includes a 39-item battery on structural social capital and a 21-item battery on cognitive social capital, field tested as a stand-alone instrument or as one that can be incorporated into ongoing survey research[4] (in the present study, it is JFPM; and (3) an organisational profile designed to delineate the relationships and networks that exist among formal and informal institutions, integrating semi-structured interview data with a scoring system for assessing organisational capacity and sustainability.

The fieldwork was carried out in two phases.[5] In the first phase, census method was followed to carry out household survey of all the members of the VFC through a structured questionnaire with 92 questions. The questionnaire had three sections. The first section included the general information of the members of the household. The second section included questions related to the awareness, participation in the JFPM programme, engagement of the VFC in and its impact on the conservation and developmental activities. The third section included questions on social capital (structural and cognitive) and interpersonal engagements. Detailed personal interviews and focus group discussions (FGDs) were held with the target group, i.e., the Korama families in the village. Only those details relevant to the present research theme are given in this paper.

Ethnographic method was followed in the second phase to observe and record major incidents in the village that influenced the process of forest conservation and development. Personal interviews with the members of the VFC, who have played a major role in the JFPM process, NGO activists and forest officials, were conducted. The FGD, mainly, was held with VFC managing committee members, landless labourers and members of various associations. The research strategy focused on the data collection for all direct stakeholders of the programme, indirect or external stakeholders like the nodal

3. Krishna, A and Elizabeth Shrader (1999). p. 6 and pp. 7-8. (html version).

4. Structural social capital includes horizontal organisational structure; collective/transparent decision-making process; accountability of leaders and practices of collective action and responsibility. Cognitive social capital includes values; trust; solidarity; reciprocity; social norms; behaviour and attitudes. This distinction is made by Norman Uphoff. For more details see Uphoff (1999).

5. See Krishna and Shrader, 1999: 12 (html version). They suggest that the household survey is most effective when it follows the completion of the community profile; however, household interviews may be conducted prior to community interviews, particularly in situations of limited time or resources.

agency/ies (Central or state governments), funding agencies, NGOs, etc. Minutes books of the VFCs were also referred to record the proceedings of and decisions taken during the general body and managing committee meetings. The units of analysis in the present study are both the household and the community, and the variables of interest related to social capital that may be created or accessed by individuals, households or local level institutions.

Understanding from the Literature

This section provides a theoretical framework for the study. A brief discussion on the concept of social capital and JFPM provides an insight to the reader on the larger issue of conflict dynamics and resolution mechanism in multi-communities' projects.

Social Capital

Social capital is a sociological concept used in general to refer to connections within and between social networks (Bourdieu, 1983/1986; Coleman, 1988/1990; Lin Nan, 2001; Portes and Landolt, 1996; Putnam, 1993; Sobel, 1994). Though there are a variety of related definitions, they tend to share the core idea that social networks have value and that different types of social contacts affect the productivity of individuals and groups (Bourdieu, 1983/1986; Fine, 2001; Blomkvist and Swain, 2001; Coleman, 1988/1990; Falk and Kilpatrick, 1999; Putnam, 1993; Sobel, 1994 and World Bank, 2000). It has to be noted that social capital may be not always be invested towards positive ends (Field, 2008; Portes and Landolt, 1996; Rudolph, 2000). We would like to focus on that aspect of social capital for the purpose of the present study within the framework of James S. Coleman's concept of 'social capital'.

Social capital can be understood in a narrow sense as a set of horizontal associations between people, consisting of social networks and associated norms that have an effect on the productivity and wellbeing of the community. In its broader sense social capital includes both positive and negative aspects, consisting of vertical as well as horizontal associations between the people and includes behaviour within and among organisations. Social structure according to Gupta (2005) is a patterned and relatively stable arrangement of roles and statuses found within societies and social institutions. The idea of social structure points out the way in which societies, and institutions within them, exhibit predictable patterns of organisation, activity and social interaction.[6]

Social capital is not a single entity, but a variety of different entities having two characteristics in common: they consist of some aspect of social structure,

---

6. See Gupta (2005). A detailed definition of social structure can be found in Lin (2001): 22.

and facilitate certain actions of the individuals who are within the structure. Like other forms of capital, social capital is productive, making possible the attainment of certain ends (conflict resolution) that would not be attainable in its absence (Coleman, 1990: 302).[7] Social capital according to Coleman has two basic elements: (a) an aspect of the social structure facilitating certain actions (conflict resolution) within the social structure; (b) social capital is the resources, real or potential, gained from relationships, which in turn facilitate the action of individual actors (leadership).

Joint Forest Management

JFM is a forest management strategy under which the Forest Department and the local community enter into an agreement to jointly protect and manage forestlands adjoining villages and share responsibilities and benefits. The village community is represented through an institution, specifically formed for this purpose. This institution is called by varied nomenclatures in different states, commonly referred to as forest protection committees or FPCs. It is called van samrakshan samitis (VSS) in Andhra Pradesh and Orissa, VFCs in Karnataka and Madhya Pradesh, hill resource management societies (HRMS) in Haryana. Some states have also made provisions for Panchayati Raj Institutions (PRIs) to enter into JFM agreements with the Forest Department.[8]

JFM was a result of a opinion that forest management programme needs to be reoriented in such a manner that they respond to the needs of the local community and that the forest produce like fuel wood, minor forest produce, timber, etc., are made available to the village community preferentially, in addition to a share in the proceeds.[9] Early experiments on JFM were conducted in the 1970s in parts of West Bengal and Haryana by local Officials, who unsuccessfully tried to keep the people out of forests through coercive measures like fines and arrests. It was also realised that active and willing participation of the forest fringe communities is necessary for any forest regeneration programme to be successful. Participation of local communities with little or no incentives would also not be effective without being assured of direct benefits and sufficient authority in managing forest resources. The village communities get a greater access to a number of non- timber forest products (NTFPs) and a share in the revenue arising from the sale of timber in return for their responsibility for the protection of forest from fire, grazing and illicit harvesting, under the agreement on JFM.

7. Harriss (2001): 19.
8. As cited in the Indian Institute of Forest Management (IIFM, Bhopal) digital databank. It can be reached at *http:/ www.iifm.org/databank/jfm/jfm.html*
9. As stated in the Preamble of the Karnataka Joint Forest Planning and Management Resolution, 1993.

JFM was introduced through a resolution on June 1st 1990, [No. 6.2 1/89-F.P.] to implement the National Forest Policy, 1988 by the Government of India. JFM is a property rights institution, which entails not only usufruct sharing with the community, but also assigns decision-making rights over resource use to the community. Latest report of the World Bank (2006) puts the number of JFM communities under collaborative arrangement in India at around 85,000 villages, covering 27 states and 17.3 million hectares with a national average of 217 hectares per JFM committee and representing 8.4 million families. At an assumed average of four people per family, 34 million people form part of the JFM programme in the country. Andhra Pradesh, Chhattisgarh, Madhya Pradesh and Maharashtra account for two-thirds of all forest cover under JFM and 39 per cent of all JFM Committees. Tribal families form one-third of all families in the JFM programme. Though these numbers represent a remarkable success by many measures, it should be viewed with caution. Dhariam's (2005)[10] study in Maharashtra suggests that many JFM Committees may not be fully operational or engaged with the Forest Department. Out of the 10,420 JFM committees covering 65 per cent of the villages in and around the forests of Maharashtra state, only 1613 villages are actively operating under the FDA.

Joint Forest Management in Karnataka

Lele (2003) notes that after the Central government's path breaking circular of 1990, the Karnataka government set up a Consultative Committee on People's Participation in forest management with NGO representatives. The Government Order (GO) on JFPM in 1993 was a result of the discussions of the Consultative Committee and parallel discussions of British funding agency (ODA) now DFID, the forest department and the NGOs.[11] JFM in Karnataka was initiated through Government of Karnataka Notification No. AHFF/232/FAP/83 dated 12-04-1993,[12] in response to:

1. Ref: 1.G.O. No. FFD/75/FAP/83, dated 23-01-1986.
2. Letter No. 6-21/89-FP, dated 01-06-1990 from the Secretary to the Government of India, Ministry of Environment and Forests, New Delhi.
3. Correspondence ending with letter No. B2CR 100/MIS/90-91 dated 29-01-1990 from the Principal Chief Conservator of Forests, Bangalore.

---

10. As cited in World Bank (2006): 11-13.
11. See Lele (2003).
12. Excerpts from the original order AHFF/232/FAP/83 by the Government of Karnataka.

In accordance with the policy guidelines, the Government of Karnataka introduced the JFPM programme, a comprehensive scheme for people's participation in planning, development, protection and regeneration of degraded forests and other government wastelands degraded due to biotic pressure. Those forest areas and fringes predominantly inhabited by tribal population and are traditionally dependent on the forest resources for their livelihood and cultural identity can also be brought under JFPM. With a view to involve people in the programme, VFCs were constituted for every village or group of village through Government Order No. AHFF/232/FAP/86 dated 12-04-1993, superseding the earlier Government Order No. FFD/75/FA/83 dated 23-01-1986. It also created the Village Forest Development Fund (VFDF), which is usually utilised as seed capital for income generating activities at the village level. The process of establishing VFCs starts with the identification of some villagers called 'promoters' or 'facilitators' who approach the concerned range forest officer and declare their intention to participate in the programme along with a list of beneficiaries. Accordingly, the general body of the committee would consist of these and other beneficiaries—any person belonging to the village or selected group of villages and interested in forest development and conservation, by registering his/her name with the secretary of the VFC upon payment of a sum of Rs.2/- per member belonging to the general category and Re.1/- per member belonging to SCs/STs.

The original order restricted the representation to one person from each family, which was later raised to two members from each family. VFCs shall be registered as associations under Karnataka Societies Registration Act, 1960 by the respective Deputy Conservator of Forests (DCF). The VFCs would thereafter sign a legal memorandum of understanding (MoU) with the Karnataka Forest Department to be represented by the DCF (Territorial) having jurisdiction over the areas. Later, bye-laws governing its duties, responsibilities, powers and jurisdiction are given, which would also indicate the village-level boundaries for all legal and administrative purposes. In addition to the functions of the VFCs that bring them financial benefits, the plan specifies as to how the community should interact with forest areas under their jurisdiction on a sustainable basis.

The Karnataka Joint Forest Planning and Management Resolution, 1993 also prescribes sharing of forest produces or proceeds derived from the sale of forest produce at the notified (concessional) rates and through public auctions. Though the entire cost on raising and maintaining plantations in such lands shall be borne by the Forest Department, only 50 per cent goes to the Government, 25 per cent goes to the beneficiaries through VFCs as per rules prescribed for the purpose and 25 per cent to the VFDF, operated by the VFCs as per the rules framed by the Government for the purpose. A unique 'Tree

Patta Scheme' (GO No. FEE 50 FAP 2000, dated: 19-6-2002) for planting trees on roadsides, canal sides and other similar places is a special feature in the Karnataka JFPM. Another similar scheme is also provided to the VFCs to plant trees on land belonging to institutions, communities and Government departments other than the Revenue Department. There is also a provision for sharing the benefits from the above schemes as per the ratio 50 per cent, 25 per cent and 25 per cent, between the Government, the VFC and the VFDF respectively.

Social Capital Approach to Conflict Resolution in the Joint Forest Management Programme

For the purpose of our study, we define 'conflict' in terms of the existence of scarce goods (limited benefits). In the words of Coser (1956: 201)[13] conflict can be defined as the existence of an excess of claimants (four hamlets) over opportunities for adequate reward (development grants) within a social structure JFPM committee. In simple terms, conflicts exist when two (or more) parties belonging to the same organisation exchange behaviours that symbolise opposition.[14] Conflict resolution is a range of methods for alleviating or eliminating sources of conflict. In other words it is a process of resolving a dispute or disagreements.

The concept of social capital has been applied to assess the performance of many institutions including JFM. Evaluation studies have shown that programmes like JFM cannot be successful without the complete participation of the local communities and careful understanding of the issues related to social organisation. This has created an opening for applying the social capital framework to analyse and understand the functioning and performance of JFM and community's participation. Kolavalli (1995) was among the first to link JFM and social capital. He notes, 'JFM also provides a means for the state to create social capital utilising labour in rural areas.'[15]

D'Silva and Pai[16] examined whether social capital is important for successful development outcomes at the grassroot in forest protection and watershed development. They argue that social capital is indeed crucial for successful functioning of participatory programmes like JFM and watershed development. Such participatory programmes require a partnership between community institutions like the VSS and self-help groups (SHGs) and the local officials for sustainable management of forests, water and other resources

13. As cited in Beals and Siegel (1967): 18.
14. Ibid.
15. See Kolavalli (1995).
16. D'Silva and Pai (2003).

based upon trust, dialogue and mutually defined rights and responsibilities. The study suggests that collective action is successful where an underlying tendency for united action already exists in a community based on cultural values, common identity, a tradition of participation and shared historical experiences. Three important features—social cohesion, effective leadership and supportive role of the local bureaucracy, with a shared understanding of participation, determine success or failure of programmes like JFM and watershed development.

According to Gupta (2005) while institutions set the rules of the game, the organisation (social capital or rather lack of it) are the players who play by such rules. The interaction between the institution, which may be formal (law, statutes and administrative procedures) or informal (norms, customs, conventions) and the organisation shapes institutional changes. The institutions are the broad legal framework, the constitution and the informal rules, while, the organisations are the forest committees, the local government (*panchayats*) and the forest bureaucracy at the field level. He argues that social capital reduces the transaction costs of JFM and increases the adherence to rules in use in the forest committees and therefore has a positive effect on the condition of the forest resources. Social capital is measured by Gupta at three levels—individual, local and societal. He analyses the programme at three level: (a) operational level, which involves the assessment of social capital and community-level activities; (b) complementary level, which examines the linkages between local government and forest committees; and (c) collective choice level, which assesses the Forest Department's monitoring.

Gupta's core argument is that there is a cost attached to every activity in the JFM programme like monitoring and conserving of forests by both the forest community and the forest department. He attempts to prove that these cost factors influence the programme positively or negatively depending on the role of the various players involved. It is here that he locates the importance of social capital as a concept based on 'trust', 'norms' and 'networks' existing in the society, which can bring down these costs. i.e., social capital minimises the operational costs of JFM. Social capital optimises the benefits with a collective choice approach, simultaneously reducing the operational costs for both the players.

Gupta also suggests that the following factors contribute to the successful functioning of forest committees:[17] These factors contribute to lesser conflicts in the organisation or better management of conflict resolution.

- Easier decision-making because of a homogeneous community.

17. Gupta (2005): 142.

- High level of interaction within the community because of a small number of households concentrated in a single settlement or adjunct settlements.
- Homogeneity plus a high level of interaction ensures a high level of social capital.
- Political homogeneity in the community.
- An active NGO, which has helped in organising the community for collective action in various arenas, forestry being one of them.
- A leadership that is easily identified and active.

It is in these two contexts in the larger framework of Coleman's concept of social capital, that we view our present case study. We pick the ideas of social cohesion[18] and effective leadership from D'Silva and Pai (2003) and the reduction of operational cost (accruing out of conflict in the association) through the concept of social capita based on 'trust', 'norms' and 'networks' from Gupta (2005).

Case Study

This particular JFPM committee was funded under two plans; first, from grants under the ODA and later under the NAP through the FDA.

Under financial assistance in the second phase of the JFPM programme funded by the NAP and implemented by the FDA, a certain amount (approximately Rs. 1.5 lakh) was provided under development grant. The managing committee on the advice of the forest officials decided to construct a community hall. But since this revenue village consists of four hamlets, it was very difficult for the managing committee to decide on the exact site for the construction. Two hamlets were shortlisted Doddabylu and Kanagodu as these are the hamlets that have the largest stakes in forest conservation activities. On 31st March 2007, meetings with the villagers were conducted in Kanagodu and Doddabylu separately to understand the demands of the members and get the feedback. During these meetings, the feedback received was that Doddabylu had not received any developmental grants so far from any government agencies and hence they deserved to be allotted the community centre.

The very next day (1st April, 2007) in the meeting at Kanagodu, two issues were discussed: one, construction of the community hall and two, minor civil works related to the village tank. Construction of community hall was ruled

18. Social cohesion is the sum over a population of individuals' willingness to cooperate with each other without coercion in the complex set of social relations needed by individuals to complete their life courses. See Stanley (2003): 09.

out as there was one already existing (under grants from a different scheme) and since the *gram panchayat* had assured to provide some money for the tank renovation, this issue too was ruled out.

The actual plan of the VFC managing committee was to construct a small office for VFC activities with a builtup area of 10'×10'. But two women's SHGs in Doddabylu prevailed upon the VFC president to construct a community hall, which would provide them with a place for meeting. This grant was under entry point activity (EPA) and it was for the VFC managing committee to decide what best to do with the money, and with many VFC's in the area having constructed community halls, the managing committee also decided to spend the money on community hall at Doddabylu. Though there were discussions on spending some money on tank renovation, there was a problem of selecting one tank out of the four in different hamlets. Hence, this idea was dropped too and finally settled for the construction of the community hall.

After the resolution being passed that the community hall would be constructed in Doddabylu, it was very difficult to find a site for the hall. The planning and construction of the community hall in Doddabylu had its own share of problems. Though the community hall was allotted to Doddabylu, not many in the village were willing to part away with their land and those who did, their land and location was not acceptable to all. Around 15 members from Station-Balegundi, 3 men and 12 women attended the deliberations related to the community hall. They wanted the community hall to be constructed in their hamlet or at least in a centrally located place which would be equally accessible to both the hamlets. They were even ready to settle for a single room office for conducting meetings. There were bitter arguments among the delegates. No amount of persuasion worked, even the assurance by the VFC president that the next developmental activity would be allotted to their hamlet. Women delegates from Station-Balegundi came out of the meeting disappointed.

Though there was no opposition to the project from people in Kanagodu, as a goodwill gesture, the VFC granted Rs. 8,000 from EPA to the SC colony in Kanagodu in the month of May 2007 to construct a small temple for their deity Chowdamma.

The people of Doddabylu were asked to look for a suitable site for the community hall. Two sites were selected, but the VFC expressed the inability to pay for the site due to limited funding. The owner (Mrs. Gowramma) of the land was not willing initially to part with her *khata* land valued at around Rs. 12,000/- and when she was persuaded upon, that her name would remain so, ever after her and when no other villager was willing to give the land, with fear that the fund would go to some other village, the whole family decided to part away with their land. The lady also agreed to the demand for the benefit of all the villagers. She was paid a goodwill amount of Rs. 4000/-, which was collected from the villagers. The lady was not satisfied with the money given to her in exchange for the land; she demanded that a small temple be built for their family deity located just behind the plot, where the community hall was to be constructed. The hall has already been constructed and the last Ganesha festival (2007) was celebrated there.

The Contending Interests

In this case study, we find that there were two incidents of contending interests within the organisation. In the first of the contending interests, whether to construct a community hall or not and where the hall needed to be constructed (whether Doddabylu or Kanagodu) was relatively easy to solve. When it was decided through social cohesion that the community hall would be constructed in Doddabylu, the actual conflict of interest emerged. There was a conflict of interest between the members of the VFC belonging to two different hamlets; Doddabylu and Station-Balegundi. Out of the four hamlets, two[19] were closer to the conservation sites, and hence they were the most active when it came to the conservation and management activities. Two houses under the JFPM programme were constructed in Station-Balegundi, despite opposition from members of other hamlets. This was done since the leadership of the VFC felt that cooperation from the members of that particular hamlet was very crucial as far as the conservation activities in the VFC plantations were concerned. Otherwise, the villagers would not cooperate in any activities.

Doddabylu hamlet had not received any development aid under the JFPM programme, yet the members from Station-Balegundi wanted the community hall in their hamlet. There were no restrictions placed on them for organising functions in the community hall. The only rationale for their demand was that they had more member households, so they remained adamant. A multiplicity of instruments sourced from social capital, cohesion, social norms, trust, effective leadership and networks came into play and the conflict that arose out of the idea of construction of community hall was resolved. The supportive role of the local forest officials, who were very supportive of the managing committee decision, was also crucial in resolving the conflict.

The story of the actual construction of the community hall was also interesting. The members of Doddabylu hamlet were so enthusiastic about the construction that nearly 25 per cent of the expenditure came from voluntary

19. Kanagodu and Station-Balegundi.

labour. Network among various SHGs and youth associations in the hamlet was utilised to maximum extent, despite the fact that members from Station-Balegundi were not in any way cooperative. The financial grant that was released was insufficient and was not that was actually assured. Now there is a sense of belonging among the villagers with the community hall, and many functions and group meetings are held there.

Conclusion

The case study suggests that the concept of social capital based on 'trust', 'norms', 'values', 'networks', 'social cohesion' and 'effective leadership' have indeed played a very important role in resolving the conflict that arose out of the process of benefit sharing. To conclude, we discuss briefly how different elements of social capital contributed to the resolution of the conflict.

First and foremost, it was the prevailing 'social norm' (Gupta, 2005) that the benefits have to be shared (to some extent, if not proportionately), that helped the leadership decide on constructing the community hall at Doddabylu with consent from a particular community (Madigas) in Kanagodu.

Second, SHGs and youth associations from Doddabylu used their 'networks' (Gupta, 2005) and thus prevailed upon the leadership to construct a community hall instead of the original plan of constructing an office for the VFC. These group-based 'networks' helped various associations to join hands and contribute to the construction of the community hall through voluntary labour and thus reduced the cost of operations.

Third, 'effective leadership' (D'Silva and Pai, 2003) proved to be very crucial in the whole process. These negotiations for resolution of the conflicts would not have been successful had it not been for the mutual 'trust' (Gupta, 2005) between the leader and members belonging to different hamlets. While the leadership was successful in constructing the hall in Doddabylu, opposition of members from Kanagodu was nullified through the 'trust' in the leadership and the 'network' put to use by the leadership with the forest officials, to release some grants (though minor) for villagers from Kanagodu. The leadership also used the tool of 'social cohesion' to the maximum extent by roping in the members from Station-Balegundi by pacifying the agitated members and assuring that the forthcoming development grant will be utilised for their hamlet. This process also suggests that there was some amount of 'trust' in the leadership that forced the members from Station-Balegundi give in and agree to the construction of the community hall in Doddabylu.

Fourth, the cost of conflict to the whole JFPM programme could have been very high resulting from the non-cooperation from one particular hamlet. The first three concluding remarks amply suggest that through social capital based on 'trust', 'networks' 'norms' and 'leadership', the cost of operations was reduced (Gupta, 2005).

Finally, one cannot deny the crucial 'supportive role of the local bureaucracy' (D'Silva and Pai, 2003) or the 'forest officials'. Their cooperation with the VFC managing committee and going along with the decisions taken during the meetings with members of the various hamlets did help in the smooth grounding of the development and conservation project.

References

Bahuguna, V.K. (2003). "Joint Forest Management in India: Current Status and Research Needs", in *A Decade of Joint Forest Management, Retrospection and Introspection: Proceedings of the International Workshop on Joint Forest Management*, MoEF and Ford Foundation, June 19-20, 2000.

Ballabh, V. and Katar Singh (1988). "Managing Forests Through People's Institutions: A Case Study of Van Panchayats in Uttar Pradesh Hills", *Indian Journal of Agricultural Economics* 43(3): 296-304.

Beals, Alan R. and Bernard J. Seigel (1967). "Organisations and Conflict", in *Divisiveness and Social Conflict An Anthropological Approach*. Stanford: Stanford University Press and OUP: Mumbai.

Blomkvist, Hans and Ashok Swain (2001). "Investigating Democracy and Social Capital in India", *Economic and Political Weekly* 36(8): 639-643.

Bourdieu, Pierre (1983/1986). "The Forms of Capital", in J.G. Richardson (ed.), *Handbook of Theory and Research for the Sociology of Education*. New York: Greenwood Press. pp.241-258.

Castle, Emery N. (2002). "Social Capital: An Interdisciplinary Concept", *Rural Sociology* 67(3): 331-349.

Coleman, James S. (1990). *Foundations of Social Theory*. Cambridge, MA.: Belknap Press.

———. (1988). "Social Capital in the Creation of Human Capital", *The American journal of Sociology* 94: S95-S120, Supplement: *Organisations and Institutions: Sociological and Economic Approaches to the Analysis of Social Structure*.

———. (1980). *Individual Interests and Collective Action- Selected Essays*. Cambridge: Cambridge University Press.

D'Silva, Emmanuel and Sudha Pai (2003). "Social Capital and Collective Action- Development Outcomes in Forest Protection and Watershed Development", *Economic and Political Weekly* 38(14): 1404-1415.

D'Souza, Ozmond Roshan (2005). "Inter Associational Conflicts and Cooperation: Implications for Development: A Case study of two Villages in Karnataka", in Bishnu C. Barik and Umesh C. Sahoo (eds.), *Panchayati Raj Institutions and Rural Development*, New Delhi: Rawat Publications. pp. 261-284.

Dahal, Ganga Ram and Krishna Prasad Adhikari (2008). "Bridging, Linking and Bonding Social Capital in Collective Action- The Case of Kalahan Forest Reserve in the Philippines", *CAPRI Working Paper* No. 79, Washington: International Food Policy Research Institute.

Dasgupta, Partha and Ismail Serageldin (1994). *Social Capital: A Multifaceted Perspective*. Washington, D.C.: World Bank.

Dhariam, M. (2005). "JFM for Community-Based Natural Resource Management: Reflections", *Information Bulletin on Participatory Forest Management*, Resource Unit for Participatory Forestry, Delhi.

Falk, Ian and Sue Kilpatrick (1999). *What is Social Capital? A Study of Interaction in a Rural Community*. Australia: University of Tasmania.

Field, John (2008). *Social Capital*. New York: Routledge.

Fine, Ben (2001). "Social Capital and the Realm of Intellect", *Economic and Political Weekly* 36(08): 741-745.

Forest Department (2004). *Shimoga Vrutthada Varshika Varadhi 2003-04* (01-04-2003 rindha 31-12-2003 ravarage) Shimoga Aranya Vruttha, Karnataka Aranya Ilakhe. Shimoga.

———. (2004). *Stoola Tippani (2003-2004) December 2003 ra varage*, Shimoga Vruttha, Aranya Ilakhe, Karnataka Sarakara.

———. (2007). *Annual Administrative Report 2006-07*. Shimoga: Shimoga Forest Circle.

Government of Karnataka (1995). *Forestry in Karnataka*. Bangalore: Karnataka Forest Department.

———. (1996). *Palgondere Palu* (Share in benefit only on participation), Government Order on Joint Forest Planning and Management. Karnataka: Karnataka Forest Department.

———. (1996). *Government Order Nos. A.H.F.S. 232 F.A.P. 86, Bangalore dated 12-04-1993 and F.E.E 94 F.A.C., Bangalore dated 16-12-1996. Joint Forest Planning and Management*, (Kannada), Bangalore: Karnataka Forest Department, Aranya Bhavan.

———. (2001). Joint Forest Planning and Management- Guidelines on the Formation of Village Forest Committees (Kannada), Bangalore: Karnataka Forest Department.

———. (2002). *Care and Share*, Revised Government Order on Joint Forest Planning and Management in Karnataka, Bangalore: Karnataka Forest Department.

———. (2002). *Implementation Manual.* (Kannada) National Afforestation Programme, Forest Development Agency, Divisional Forest Office, Sagar, Shimoga.

———. (2007). *Annual Report*, Karnataka Forest Department, Bangalore.

Gupta, Krishna (2005). *The Political Economy of Forest Management: Importance of Institutions and Social Capital.* New Delhi: Allied Publishers Pvt. Ltd.

Harriss, John (2001). *Depoliticising Development- The World Bank and Social Capital*. New Delhi: Left Word Books.

Hill, I and D. Shields (1998). "Incentives for Joint Forest Management in India- Analytical Methods and Case Studies", *World Bank Technical Paper* No. 394, pps. vi-xi, 1-4 and 29-38.

Judkins, Benjamin N. and Stephen E. Reynolds (2009). "Globalisation, Social Capital and Networked Violence: The Role of Values", in Ibrahim A. Karawan, Wayne McCormack and Stephen E. Reynolds (eds.), *Values and Violence Intangible Aspects of Terrorism*, Netherlands. Springer. pp. 137-150

Kameshwari, V.L.V (2001). "Collective Management of Forest Resources in India: Lessons for Policy Formulation and Stakeholders Involvement", Paper submitted to the 9th Biennial Conference of International Association for the Study of Common Property (IASCP), Victoria Falls, Zimbabwe.

Khare, Aravind, James Mayers and Elaine Morrison (ed.) (2000). *Joint Forest Management: Policy, Practice and Prospects*. London: World Wide Fund for Nature-India and International Institute for Environment and Development.

Kolavalli, Shashi (1995). "Joint Forest Management- Superior Property Rights?", *Economic and Political Weekly* 30(30): 1933-1938.

Krishna, A. and Elizabeth Shrader (1999). "Social Capital Assessment Tool", Paper prepared for the Conference on *Social Capital and Poverty Reduction*, Washington, DC: The World Bank, 22-24 June. Html version of the file *http://wbln0018.worldbank.org/external/lac/lac.nsf/0/d2d929b5fff4b555852567ee00041ad/$FILE/Drft0623.doc*

Lele, Sharachchandra, A.K. Kiran Kumar and Pravin Shivashankar (2005). "Joint Forest Planning and Management in the Eastern Plains Region of Karnataka: A Rapid Assessment", *Technical Report*, Bangalore: CISED.

Lele, Sharchchandra (2003). "Participatory Forest Management in Karnataka At Crossroads", *Community Forestry* 2(4): 4-11.

Lin, Nan (2001). *Social Capital: A Theory of Social Structure and Action*. New York: Cambridge University Press.

Mohan, Giles and John Mohan (2002). "Placing Social Capital", *Progress in Human Geography* 26(2): 200. Also see page 6 (html version) of Krishna and Shrader (1999).

MoEF (Ministry of Environment and Forests, Government of India) (2000). *Guidelines for Strengthening of Joint Forest Management (JFM) Programme*, Forest Protection Division, No 22-8/2000-JFM (FPD), dated 21 February, 2000.

Naik, Gopal (1997). "Joint Forest Management—Factors Influencing House Hold Participation", *Economic and Political Weekly* 32(48): 3084-3089.

Norman, Uphoff (1999). "Understanding Social Capital: Learning from the Analysis and Experience of Participation", in Partha Dasgupta and Ismail Serageldin (ed.), *Social Capital: A Multifaceted Perspective*, World Bank, Washington D.C., p. 217.

Poffenberger, Mark and Betsy McGean (eds.) (1996). *Village Voices, Forest Choices: Joint Forest Management in India*. New Delhi: OUP.

Portes, Alejandro and Patricia Landolt (1996). "Unsolved Mysteries: The Tocqueville Files II- The Downside of Social Capital", *The American Prospect* 07(26), May 1 to June 1.

Puliani, Sathpal (ed.) (2000). *The Karnataka Forest Manual*, Karnataka Law Journal Publications, Bangalore: Law Publishers.

Putnam, Robert D. (2000). *Bowling Alone: The Collapse and Revival of American Community*. New York: Simon and Schuster.

Putnam, Robert D., Robert Leonardi and Roffaella Y. Navetti (1993). *Making Democracy Work- Civic Traditions in Modern Italy*. Princeton: Princeton University Press.

Rajpraveen, C.M (2000). "Joint Forest Management: Conceptual and Theoretical Issues", *Man and Development* 22(3): 130.

Ravindranath, N.H, K.S. Murali and K.C. Malhotra (eds.) (2000). *Joint Forest Management and Community Forestry in India: An Ecological and Institutional Perspective*. New Delhi: Oxford and IBH Publishing Co. Pvt. Ltd.

Ravindranath, N.H and P. Sudha (eds.) (2004). *Joint Forest Management in India: Spread, Performance and Impact*. Hyderabad: Universities Press.

Rudolph, Susanne Hoeber (2000). "Civil Society and the Realm of Freedom", *Economic and Political Weekly* 35(19): 1762-1769.

Sarker, Debnarayan (200). "Joint Forest Management: Critical Issues", *Economic and Political Weekly* 44(04): 15-17.

Sabel, Joel (1994). "Learning by Monitoring: The Institutions of Economic Development", in Smelser and Swedverd, *Handbook of Economic Sociology*. Princeton: Princeton University Press.

Sobel, Joel (2002). "Can We Trust Social Capital?", *Journal of Economic Literature* 40(1): 139-154.

Stanley, Dick (2003). "What do We know about Social Cohesion: The Research Perspective of the Federal Government's Social Cohesion Research Network", *Canadian Journal of Sociology* 28(01): 5-17. Special Issue on Social Cohesion in Canada.

Sundar, Nandini (2000a). "Unpacking the 'Joint' in Joint Forest Management", *Development and Change* 31: 255-279.

Sundar, Nandini, Roger Jeffrey and Neil Thin (2001). *Branching Out: A Comparative Study of Joint Forest Management in India*. New Delhi: Oxford University Press.

World Bank (2000). "What is Social Capital?", it can be reached at *www.worldbank.org/poverty/scapital/*

———. (2006). *India: Unlocking Opportunities for Forest Dependent People*, Agriculture and Rural Development Sector Unit, South Asia Region, New Delhi: Oxford University Press, pps. xiii-xxvi, 11-13 and 78-83.

15

Joint Forest Management-led Socioeconomic and Environmental Change in Orissa

R.K. PANDA and C.R. DAS

Background

A large number of people living in and around forest areas depend heavily on forests for their livelihood. As per one estimate 60 million people living in the rain forests of Latin America, Southeast Asia and West Africa depend heavily on forest; 350 million people living in, or next to dense forests depend on forests for sustenance or income; and 1.2 billion people in developing countries use trees on farms to generate food and cash. (DFID, 2002). The Forestry Department of the Food and Agriculture Organization of the United Nations (UNFAO) with the support of the UK's Department for International Development (DFID) in an interagency forum (2001), on the role of forestry in poverty alleviation, declared that forests and trees can help to eliminate poverty and promote food security. Forests can provide commercial opportunities and employment for the poor and it is central to the development of good local governance. A people-centered approach can further increase the impact of forests and trees in reducing poverty. What is needed is the removal of barriers that prevent forests and trees from contributing to the livelihoods of the poor as well as support for emerging opportunities. In recent years a debate is intensifying amongst the planners, policymakers, conservationists, forest officials, researchers and environmentalists in India (and elsewhere in the world) as to how to reconcile the seemingly incompatible goals of conserving forests, meeting market demand and promoting broad-based sustainable development that reduces rural poverty. Because even today, while on the one hand, the government continues to claim ownership over the forests—extensively depriving the poor communities for securing their traditional benefits of food subsistence, income from forests, on the other hand the government has begun to involve forest dependent communities for protection, regeneration and conservation of forestry.

Forest Resources in India

India is endowed with forest resources, rich in diverse flora and fauna. The forest types vary from tropical rain forest in northeastern India, Western Ghats and Andaman and Nicobar Islands to desert and thorn forests in Gujarat and Rajasthan, rich mangrove forests in West Bengal, Orissa, Andhra Pradesh and Andaman and Nicobar Islands to dry alpine forests in Western Himalayas. India has approximately 7 per cent of total mangrove forests of the world and the seventh largest country in the world covering an area of 32,87,000 km^2 lying entirely in the northern hemisphere. State-wise occurrence of different forest types in India is shown in Table 15.1. Tropical moist deciduous forest and tropical dry deciduous forest are the major forests occurring in India and their proportionate share to the total forest types is 65.6 per cent.

Ownership of Forest Resources in India

On the basis of the legal status, forests are classified into: (i) reserved forest; (ii) protected forest; and (iii) unclassified forest. The proportionate shares of the reserved, protected and unclassified forests are 53, 30 and 17 per cent respectively. India has a recorded forest area of 76.52 million hectares that is 23.28 per cent of its total geographical area. The ownership of this forest area largely remains under government sector. It is estimated that 92.47 per cent of the recorded forest area is owned by the forest departments, 3.18 per cent by other government departments, 2.45 per cent by corporate bodies, and 1.91 per cent by others (Sagereiya, 1994). Ownership of forest resources in India is shown in Table 15.2.

Depletion of Forest Coverage

According to an estimate of FAO, India lost 3.4 million hectares of forest area between 1951 and 1972, which worked out to an annual denudation of 1,55,000 hectares. But the latest satellite imagery studies point out that the country is currently losing its forest cover at the annual rate of nearly 1.3 million hectares (FAO, 1981). Between 1951 and 1972, over 72 per cent of forest area was lost to agriculture and another 17 per cent was lost to river valley projects, industries roads and communications.

Shifting Cultivation

Shifting cultivation is one of the primitive systems of raising food crops with 'slash and burn' technique. It has been in practice since long and got ingrained into the tradition and culture of most of the tribal community in the country. Northeastern states in India have more concentration of tribal people. Loss of forest cover is very acute in northeastern region, and primary reason

Table 15.1

Occurrence of Forest Types in India

Sl. No.	*Forest Type*	*Area in Sq Km*	*%*	*Occurrence*
1.	Tropical wet evergreen forest	51,249	8.0	Arunachal Pradesh, Assam, Karnataka, Kerala, Manipur, Nagaland. Tamil Nadu, Andaman & Nicobar Island and Goa.
2.	Tropical semi-evergreen forest	26,794	4.1	Assam, Gujarat, Karnataka, Kerala Maharashtra, Goa, Nagaland, Orissa, Tamil Nadu, Andaman and Nicobar Island.
3.	Tropical moist deciduous forest	236,794	37.0	Andhra Pradesh, Assam, Gujarat, MP Maharashtra, Kerala, Maharashtra, Bihar, Manipur, Meghalaya. Mizoram, Tripura, Nagaland, Orissa, Tamil Nadu, UP, Goa West Bengal, Andaman & Nicobar Island and Dadra and Nagar Haveli.
4.	Littoral and swamp forest	4,046	0.6	Andhra Pradesh, Gujarat, Maharashtra, Orissa, Tamil Nadu, West Bengal and Andaman & Nicobar Island.
5.	Tropical dry deciduous forest	186,620	28.6	Andhra Pradesh, Gujarat, Haryana, HP, Karnataka, MP, Maharashtra, Jammu and Kashmir, Orissa, Punjab, Rajasthan, UP, Tamil Nadu, West Bengal and Kerala.
6.	Tropical thorn forest	16,491	2.6	Andhra Pradesh, UP, Gujarat, Haryana, Karnataka, MP, Maharashtra, Punjab, Rajasthan, HP, and Tamil Nadu.
7.	Tropical dry evergreen forest	1,404	0.2	Andhra Pradesh and Tamil Nadu.
8.	Sub-tropical broad leafed hill forest	2,781	0.4	Assam, Maharashtra, Meghalya, and West Bengal.
9.	Sub-tropical Pine forest	42,377	6.6	Arunachal Pradesh, HP, Jammu and Kashmir, Manipur, Meghalaya, Nagaland. Sikkim and UP.
10.	Sub-tropical dry evergreen forest	12,538	2.5	HP, Jammu and Kashmir and Mizoram
11.	Mountain wet temperate forest	23,365	3.5	Arunachal Pradesh, Karnataka, Manipur, Nagaland, Sikkim and Tamil Nadu.
12.	Himalayan moist temperate forests	22,012	3.4	HP, Jammu and Kashmir and UP.
13.	Himalayan dry temperate forest	313	-	Jammu and Kashmir and HP.
14.	Sub-alpine and alpine forest	18,628	2.8	Arunchal Pradesh, HP, Jammu and Kashmir, Nagaland, Sikkim and UP.

Source: National Wasteland Development Board, 1988.

Table 15.2

Ownership of Forest Resources

Ownership Category	*Area (in Thousand Ha)*	*Per cent of Total Forest Area*	*Per cent of Total Geographical Area*
Forest land under public ownership	(73780)	(97.36)	(22.44)
• Through forest departments	64513	85.13	19.62
• Through other departments	9267	12.23	2.82
Forest land under other ownership	(2001)	(2.64)	(0.61)
• Corporate bodies	1057	1.39	0.32
• Private individuals	944	1.25	0.29
Total	75781	100.00	23.05

for this loss is shifting cultivation. Shifting cultivation is practiced in parts of the states of Andhra Pradesh, Bihar, Madhya Pradesh, Orissa, in addition to all seven northeastern states. Incidence of shifting cultivation is very heavy in Orissa and northeastern states (Arunachal Pradesh, Assam, Manipur, Meghalaya, Mizoram, Nagaland and Tripura). Traditionally shifting cultivation was practiced on long cycles with rest period ranging from 10 to 30 years. However with increase in population of shifting cultivators, cycle of shifting cultivation is reduced to below than 5 years and sometimes down to 2 years. It has resulted in severe loss to forest resources and biodiversity. It has also increased misery of shifting cultivators as the agricultural yield is falling rapidly due to decreasing productive capacity of soil in the shifting cultivation area (Forest Survey of India, 1995).

Encroachments on Forest Lands

Reliable data on encroachments on forestlands are not available. Forest Survey of India (FSI) estimated in 1987 that over 700,000 hectares of forestlands are under encroachments. By now it has increased considerably. Most of the forest area under encroachments is being used for agricultural purpose. At several places it has been observed that some encroachments began as a sort of shifting cultivation and then turned into settlement and permanent cultivation. (State Forest Report, 1995; 1991).

Grazing

Due to very high cattle population (450 million), there is severe shortage of fodder. Cattle are generally allowed to graze openly in forest areas. Even though grazing is prohibited in protected areas, 67 per cent of national parks and 83 per cent of sanctuaries have reported incidence of grazing. There are two main types of livestock pressures on forest areas. One is pressure of sedentary village livestock and small ruminants and other is pressure from migratory animals reared by traditional ethnic grazers (Gujjars and Maldharies). Some of them also live in the forest with their livestock continually affecting the vegetation therein.

Forest Fires

Fires are a major cause of injury to forest areas. It is estimated that most of the forests suffer burns annually. Most of the fires in forest of India are surface or ground fires. Crown fires seldom occur and reported from coniferous forests in Himalayas. Sometimes ground fires are also reported from forests at high altitudes. Nearly 98 per cent of fires in the country are caused by people. Recently Government of India (GoI) has realised the significance of forest fires, and initiated Centrally sponsored scheme 'Modern Forest Fire Control Methods' to support, control and management of fires in forest areas (GoI, 1995).

Joint Forest Management

The development of the joint forest management programme (JFMP) in India is based on the National Forest Policy (NFP) of 1988 which emphasised the involvement of the village communities living close to the forests in protection and development of forests and development of the degraded forestland. Besides, NFP 1988 laid emphasis on natural regeneration, preserving biodiversity and adopting management practices that would help the local people. The translation and implementation of this basic policy for protecting forests and environment and reducing the poverty level of rural masses paved the way for the emergence of JFM. The GoI, Ministry of Environment and Forests (MoEF) issued an enabling joint forest management resolution on June 1st, 1990 for the active involvement of village community living close to the forests and development of degraded forest land. The main features include the following:

It was integrated with socioeconomic realities in a multiple cultural and ethnic setup embarking on a path of productivity and prosperity through sustainable use of scarce natural resources.

In the year 1995, the resolution was slightly modified to involve women and landless households as well as usufruct sharing by the communities.

To provide legal backup to JFM committees, it was further revised on 21st february 2000. Orissa is among the 27 states to implement the JFM programme and it is the third highest state to implement the JFM programme more vigorously after Madhya Pradesh and Andhra Pradesh (Planning Commission, 2001). The village forest protection committee (FPC) is called

Vana Suraksha Samitti (VSS). In the light of the NFP 1988, JFM has been considered to be the right policy intervention for ensuring livelihood improvement of the forest dependent poor in Orissa, who are underorganised and underprivileged due to their poverty, ignorance and impoverishment. By the end of June 2004, there were 7358 VSSs in the 49 forest divisions covered under 8 forest circles in Orissa. The total forest area covered under the JFM programme stands at 8,03,807.16 hectares (see Table 15.3).

Table 15.3

Circle-wise Performance of JFMs in Orissa (As on 30.06.2004)

Sl. No	*Circles*	*No. of VSSs*	*Assigned Forest Area (in Hectares)*
1.	Angul (Angul, Athagarh, Athamallik, Cuttack, Dhenkanal, Satkosia WL)	632	72454.58
2.	STR-Baripada (Baripada, Balasore WL, Bhadrak-WL, Karanjia, Rairangapur, STR-Baripada)	702	168707.41
3.	Berhampur (Baliguda, Baudh, Berhampur, Ghumsar-N, Ghumsar-S, Paralakhemundi, Phulbani)	1235	108261.79
4.	Bhawanipatna(Bolangir-E, Bolangir-W, Kalahandi-N, Kalahandi-S, Khariar, Sunabeda-WL)	1526	116966.61
5.	Bhubaneswar (Chandaka-WL, Chilka-WL, City Forest, Khurda, Mahanadi-WL, Nayagarh, Puri-WL, Rajnagar-WL)	219	26772.65
6.	Koraput (Jeypore, Koraput, Malkangiri, Nabarangpur, Rayagada)	1292	91252.04
7.	Rourkela(Bonai, Deogarh, Keonjhar, Keonjhar-WL, Rourkela, Sundargarh)	1083	119149.41
8.	Sambalpur (Bamra-WL, Basuguda, Hirakud-WL, Rairakhol, Sambalpur-N, Sambalpur-S)	669	100332.69
	Total	7358	803807.18

Note: Names in Parenthesis are the respective forest divisions; STR = Similipal Tiger Reserve; WL = wildlife sanctuary

Source: Office of the PCCF, Orissa, Bhubaneswar (2006).

Need and Objectives of the Study

Orissa is one among the major states to implement the JFM programme and till date more than 7,000 VSSs have been formed in different forest circles in Orissa. It is the right time to assess different impacts of this programme towards the sustainable rural development in Orissa. In this background, present study analyses the socioeconomic, environmental changes in the JFM areas of Orissa.

Study Approach and Methodology

For analysing the changes due to JFM programme 16 VSS villages under Digapahandi forest range and Gundibira VSS under Surada forest range are considered. The socioeconomic and the environmental changes are assessed on the basis of a number of selected indicators. The performance of the selected indicators is outlined on the basis of primary data obtained from the people in the selected VSS areas with the help of focus group discussion (FGD) and rapid rural appraisal (RRA). The changes due to JFM programme is measured by comparing the values of the selected variables during the post-JFM period with the corresponding figures during the Pre-JFM period. In this study 9 VSSs—Pipiliguda, Narasinghagada, Gaida, Moreibadi, Raipur, Patrachudi, Harina, Kandhakhududi, Kamalpur under Digapahandi forest range and Gundibira under Surada forest range are considered.

Socioeconomic Impacts of JFM Programme

To examine the socioeconomic changes due to JFM, the indicators like incidence of small farmers, marginal farmers, agricultural labourers, forest dependent households and the type of migration etc., are analysed. The details of the socioeconomic indicators and their changes due to the present JFM programme are shown in Table 15.4.

It is revealed from Table 15.4 that during the post-VSS period, the incidence of small farmers remains constant, the incidence of marginal farmers has increased by 51.25 per cent and this has reduced to 30 per cent for agricultural labour households. This indicates that most of the agricultural labour households during the pre-VSS period have moved upward and become marginal farmers. Similarly the incidence of forest dependent households, those who are exclusively dependent on forests for their livelihood is significantly reduced and the extent of reduction is around 90 per cent. It is inferred that there has been better occupational mobility and better awareness for the protection of forests. Besides, these households are pressurised by JFM members not to be engaged in illicit timber trafficking, collection and sales activity for some better economic activity. As a result, outmigration from the VSS area to Surat textile mills has increased to the extent of 120 per cent. Similarly, livestock activities during the post-VSS period have also increased sharply. The increase in the livestock population such as cows, bullocks, goats, and sheep has increased by 66 per cent, 400 per cent, and 233.33 per cent respectively. During the pre-VSS period there were no Holstein cows in the JFM villages which are reported to be around 10 per village.

The average milk production during a month in the VSS area has increased by 66 per cent. The increase in the number of shops, artisan households and women entrepreneurship speaks about the occupational mobility of the forest dependent households. There is increasing tendency among women to be associated with microfinance activities through self-help group (SHG) programme. For instance, the JFM formed SHG comprising forest fringe women members of Gundibira VSS under Ghumsar (South) forest division have produced beautiful market-oriented seasons greeting cards and wall calendars.

Table 15.4

Socioeconomic Changes Due to JFM

Sl. No.	*Indicators*	*Pre-VSS*	*Post-VSS*	*Net Change (% Deviation)*
1.	Incidence of small farmers (% of households to the total households)/VSS	19.08	19.08	Nil
2.	Incidence of marginal farmers(% of households to the total households)/VSS	28.24	42.74	51.25
3.	Incidence of agricultural labour households (% of households to the total households)/VSS	48.85	34.35	-29.66
4.	Incidence of forest dependent households (% of households to the total households)/VSS	38.16	3.81	-90.01
5.	Migration of adult labourers (No. of persons)/VSS	25	55	120.00
6.	Migration of child labourers (No. of child labourers)/VSS	15	6	-60.00
7.	Dropout rate[1]	40.81	12.72	-68.83
8.	Low birth weight babies[2]/year (Weighted average separately calculated during pre-VSS and post-VSS)	1.5	1.0	-33.33
9.	Livestock population (in no.s/VSS)			
	No. of cows	300	500	66.66
	No. of bullocks	12	60	400.00
	No. of goats/sheep	60	200	233.33
10.	Average milk production (litres/month/VSS)	900	1500	66.66
11.	Shops (nos./VSS)	2	5	150.00
12.	Artisan households (nos./VSS)	5	10	50.00
13.	Women entrepreneurship (no. of SHGs/VSS)	0	6	-

Note : All the figures in column 3 and 4 are the averages of all the JFM villages in the study.

1. Out of the enrolled children in primary schools, a proportion of children have left school in between the education sessions. The mean dropout rate of the preceding five years to the introduction of the VSS and succeeding three years has been considered.
2. Babies weighing less than 2,500 grams during birth are considered low birth weight (LBW).

Source : Computed from the field data.

Even the human development indicators like dropout rate and the incidence of low birth weight babies have reduced. This is due to the increased awareness consequent upon the health camps and literacy campaigns periodically undertaken in the VSS areas.

Environmental Impacts of JFM

The environmental impacts of JFM are examined by analysing the indicators associated with soil conservation, moisture conservation and improved environment. Indicators like number of tube wells, number of dug wells, both type of wells remaining dry during summer, mean temperature during summer etc., have been studied to understand and assess the extent of moisture stress created by the adjoining forestry now covered under the JFM programme. It is revealed from Table 15.5 that though there has not been any increase in the mean number of tube wells per village, the number of dug wells have increased by 14.28 per cent. The number of tube wells have not increased because tube dug wells in the villages are installed by the government. It is also found that the no of tube wells and dug wells remaining dry during summer have decreased to the extent of 50 and 60 per cent respectively. Improvement in the water table is also reported to the extent of 25 per cent. Thus, overall performance with respect to moisture conservation is found satisfactory. Mean temperature during the peak summer is also reported to be declining in the microclimates adjacent to the forests covered under the JFM programme.

For assessing soil conservation, indicators like amount of sand and stone transported from VSS site to construction sites are studied. Due to the active involvement of the VSS people, such type of transportation has been checked by 50-100 per cent. As a result, soil erosion in forest areas has been prevented. This is definitely a significant outcome of VSS.

Environmental protection due to VSS efforts are analysed by studying the variables like the amount of timber transported from the forest areas to outside, number of days of forest firing during summer, number of picnics arranged at the VSS sites. It is found that during the post-VSS period, there have been around 99 per cent reduction in the amount of timber transported and 90 per cent reduction in forest firing. There has been six times increase in the number of picnic days arranged in the VSS sites.

Thus against this background the environmental promotion role of VSS is profound. Effective utilisation of minor forest produces (MFPs) is also one of the important objectives of JFM. It is reported that in the study area MFPs are not found much. However it is found that the number of households engaged in leaf plate making is sufficiently increased and the increase is as high as 400 per cent.

Table 15.5

Environmental Changes Due to JFM

Sl. No.	*Indicators*	*Pre-VSS*	*Post-VSS*	*Net Change (% Deviation)*
1.	No. of tube wells per JFM village	6	6	0
2.	No. of dug wells per JFM village	7	8	14.28
3.	No. of tube wells remaining dry during summer	2	1	-50.00
4.	No. of dug wells remaining dry during summer	8	5	-60.00
5.	Mean water table during summer	25'	20'	-25.00
7.	Mean temperature during summer	43°	40°	12.5
8.	Stone transported (no. of truckloads/week)	20	0	-100.00
9.	Red sand transported (no. of truckloads/week)	8	0	-100.00
10.	Timber transported (no. of bicycle-loads/week)	500	2	-99.6
11.	Households engaged in leaf plate making	2	10	400.00
12.	Picnics arranged at VSS site	2	15	650.00
13.	Forest firing (no. of days during summer)	20	2	-90.00

Note : All the figures in column 3 and 4 is the average of all the JFM villages considered in this study
Source : Computed from the field data, 2006.

Perceived Benefits of the JFM Programme

VSS-wise the benefits as perceived by the people is shown in Table 15.6. The different benefits as perceived by the villagers include present as well as future benefits. The whole range of benefits as summarised in the following table necessarily speaks about the sustainability symptoms.

Table 15.6

Benefits of the JFM Programme as Perceived by the Villagers

Sl. No.	*Name of the VSS*	*Project Benefits*
1.	Pipiliguda	• Forest loss prevented to the extent of 70 per cent. • Increased availability of fuel and timber in future. Increase in the income of the forest dependent households. • More employment. • Increased no of goats and sheep in the village among the forest dependent households. • More plantation consciousness.
2.	Narasinghagada	• Profuse regeneration of bamboo trees. • More number of wild animals reported in the jungle. • Complete prevention of forest fire.
3.	Gaida	• Reduction in illicit transportation of timber. • Reduction in the incidence of forest fire. • Profuse regeneration of bamboo trees.

contd...

...contd...

		• Inmigration of elephants. • Increased availability of fodder. • Increased concentration of wild animals.
4.	Moreibadi	• Increased stall feeding practice for the cattle among the villagers. • In migration of wild animals. • Flow of streams from forests. • Reduction of top soil erosion due to the soil and moisture conservation works. • Reduction in illicit stone transportation. • Reduction in the incidence of the forest dependent households. • Massive regeneration of trees. • Once extinct, reappearance of the wild species like gunduri.
5.	Raipur	• More interest to go for plantation in the private lands. • Restricted open grazing. • Introduction of crossbred cows. • Prevention of forest fire. • Additional income of the households due to increased off-farm employment opportunities.
6.	Patrachudi	• Sufficient availability of firewood. • More awareness for maintaining the forest. • Better quality of life due to the convergence of programmes like WATSAN (water and sanitation) and Swajaladhara with the JFM works.
7.	Harina	• Massive regeneration of bamboo trees. • Multiple regeneration of plants. • More availability of MFPs. • Reduction in the illicit stone transportation. • Plenty availability of firewood, timber and fodder as future benefits.
8.	Kandhakhududi	• Increased stall feeding practices. • Indigeneous cows being replaced by she-buffaloes. • Increased milk production. • Plenty availability of firewood and bamboo as future benefits.
9.	Kamalpur	• Place is first becoming a tourist place. • More inflow of tourists for arranging picnic parties in the VSS jungles. • Flow of streams from forest. • Reduction in the top soil erosion. • Inmigration of elephants. • Due to increased fodder availability in the forest, wild animals are not interfering in crop lands.
10.	Gundibira	• Future availability of fuel and timber. • Sustainability of forest resources. • Scenic beauty of the forest.

Source: Field Survey, 2006.

Problems in the Smooth Functioning of the JFM Programme

Table 15.7 shows the various limitations of the JFM programme as perceived by the villagers. The major limitations as revealed in the following table include non-adoption of Participatory Rural Appraisal (PRA) exercise while formulating the JFM programme, non-maintenance of records and registers and non-completion of micro-plan. It is reported that as the various functionaries of the executive body of the VSS are illiterates they can not provide assistance towards the preparation of micro-plan and maintaining records and registers. Another major limitation of the JFM programme is that in around 30 per cent of the VSSs, the various functionaries like president and secretary are not transparent to the member households regarding the fund flow under the VSS programme.

Table 15.7

Limitations of Joint Forest Management as Perceived by the JFM Households

Sl. No.	*Type of Limitations*	*No. of Reporting JFMs (%)*
1.	Non-completion of micro-plan	40
2.	Delay in the sanction of funds	30
3.	Non-adoption of PRA exercise	60
4.	Conflict among the households	20
5.	More interference of *gram panchayat* functionaries in JFM matters	20
6.	VSS functionaries do not provide information to the members	30
7.	Records/registers are not up-to-date.	60
8.	Forest officials dominate over JFM matters	30
9.	Minimum wage payment is violated	10
10.	Fictitious records/registers	10

Concluding Remarks

With the introduction of the JFM programme there has been sufficient preservation, protection and maintenance of forest resources, marked improvement in the socioeconomic life of the forest fringe households and positive changes in the environmental aspects. A fraction of the agricultural labour households have escaped from poverty. Meat and milk production in the VSS areas have increased. There has been satisfactory soil and moisture conservation due to the forest protection and forestation activities undertaken by the VSS. Added to such benefits, groundwater table has improved, forest firing prevented, biodiversity in VSS areas enhanced, and occupational mobility improved to a satisfactory level. Effective utilisation of MFPs is also reported. People's involvement in the protection of forests is tremendously proactive.

However issues like delays in the sanction of funds, non-transparency of the VSS functionaries, insufficient provision of training and exposure visits for the members have partially hindered the progress. Considering that there is tremendous scope for sustainable rural development of the forest fringe households through the JFM programme, all types of constraints associated with such type of governance should be removed.

References

B., Pramod and N. Murari Ballal (1992). "Sustainable Development of the Rural Poor", *Pigmy Economic Review* 38(1).

Bellamy, Carrol (2005). *The State of the World's Children*. United Nations Children's Fund.

Dash, C.R. (2007). "Joint Forest Management in Orissa: Some Issues", *Working Paper Series*. Yojana Ink, No12. /2006, Berhampur.

DFID (2002). *Sustainable Agriculture: Resource Management Key Sheet* 10. London, UK: DFID.

FAO (1981). *Forest Resources of Tropical Asia*.

Forest Survey of India (1995). *State Forest Report*. New Delhi: Ministry of Environemnt Forests, Government of India.

———. (1991). *State Forest Report*. New Delhi: Ministry of Environemnt Forests, Government of India.

GoI (1991). *State of Forest Report*, Ministry of Environment and Forest. New Delhi: GoI.

———. (1995). *State of Forest Report*, Ministry of Environment and Forest. New Delhi: GoI.

National Wasteland Development Board (1988). *Occurrence of Forest Types in India*. New Delhi: Government of India.

Office of the PCCF (2006). "Unpublished Official Records", Government of Orissa.

Panda, R.K. and J.N. Panda (2007). "Socio Economic and Environmental Changes and Joint Forest Management", in S.N. Tripathy (ed.), *Dynamics of Rural Development*. New Delhi: Discovery publishing house.

Panda, R.K. and K.C. Patnaik (2007). "Functioning of Joint Forest Management in Orissa", *Working Paper Series*, No.11./2006, Berhampur: Yojana Ink.

Planning Commission (2002). *Orissa Development Report*, Government of India.

———. (2001). *Approach Paper to Tenth Five Year Plan (2002-07)*. New Delhi: Government of India.

Sageriya, K.S. (1994). *Forests and Forestry in India*. New Delhi: National Book Trust.

Saxena, Rakesh (2004). "Governance of Forest Resources in India: A Brief Review and Identification of Current Issues", Paper presented at Eastern Zone Reflective Workshop on *Governance: Gender, Poverty and Livelihood*, held at NCDS, Bhubaneswar, 20-21 July.

16

Developing Strategies for Rehabilitation of Orans and Gauchars of Western Rajasthan

NAVEEN K. BOHRA and PRADEEP CHAUDHRY

More than two-third of India's geographical area falls within the dry land (arid, semi-arid and dry sub-humid) as per Thornthwaite climate classification and is affected by varying kinds of land degradation. The arid region of western Rajasthan is densely populated and has a high livestock to human ratio when compared to other deserts of the world (Kumar *et al.*, 2003). Overexploitation and vegetation removal are widely cited as the major cause of desertification in arid and semi-arid areas world over. In Indian arid zone, drought and famine are regular phenomena owing to very hostile climatic conditions and high temperature coupled with high population pressure. Out of every 10 years there are 3 famine years on an average. People of this region through their traditional wisdom and experience have developed strategies to face and manage these vagaries of nature. However, increasing pressure from livestock and human population along with socioeconomic development in the region has caused severe land degradation (Narayan *et al.*, 2003).

Breakdown of traditional institutions coupled with loss of collective community wisdom has resulted in the destruction of fragile ecosystem of the western Rajasthan. There is a vast cattle population in this part of the world and to get their fodder on sustainable basis, it is necessary that in almost every village there should be a provision for a portion of land to be kept aside for grazing. The culture and society of Rajasthan is very rich and Oran and Gauchar lands are the real privilege given by their culture.

Economy of the extremely arid districts in the Jodhpur revenue division of Rajasthan where the cropped area varies from 8 to 57 per cent during normal years of rainfall is largely livestock based. Animal husbandry is one of the major occupations of the people of this region. Apart from compensating farmers for poor fertility and hence low agricultural productivity, animal husbandry acts as an effective tool to fight droughts which are very frequent in this region by a way of providing additional income and other essential requirements like milk, meat, fibre and manure. From time immemorial, in order to support animal husbandry on a sustained basis in almost every village of the western Rajasthan, a substantial portion of land is kept aside specifically as grazing ground. Such grazing lands often serve as catchment areas for village pond and are commonly known as Gauchars (*gau*: cattle; *char*: grazing). In order to discourage cutting of trees some of the Gauchars have been pledged to local deities, gods, goddesses, legendary heroes etc., and are known as 'Orans'. As per common belief, word 'Oran' is derived from '*Aranya*' which means forest. Some people feel that the word 'Oran' is derived from '*Aan*' which means 'pledged'. Orans and Gauchars are the most important common property resources available in the rural areas of the western Rajasthan.

In the past, Orans and Gauchars were mainstay of livestock farming in the area. The productivity of these areas has however decreased due to heavy pressure of grazing. However, with the time, the scenario changed. The livestock population in terms of adult cattle unit (ACU) increased from 9.6 million in 1983 to 11.3 million in 2000. As a result of this, fodder deficit increased from 14.9 million tonnes in 1983 to 28.2 million tonnes by 2000. Likewise the human population, which was 17.4 million in 1991, is expected to increase to 28.4 million by 2020. Similarly water which is required for human beings and livestock consumption, irrigation and industry; is expected to increase from 5640 mcm (million cubic metres) in 1981 to 7638 mcm in 2011. All these factors heavily contributed to degradation of natural resources/ agricultural and community pasture lands (Gupta, 2003).

Such high increase in the livestock population resulted in heavy overgrazing in Orans and Gauchars much beyond their carrying capacity. Continued overgrazing of naturally growing palatable grassed legumes and shrubs like *Sevan* (*Lasiurus sindicus*), *Dhaman* (*Cenchrus ciliaris*), *Bhurat* (*Cenchrus biflorus*), *Anajn* (*Cenchrus setigenus*), *Dhamnio* (*Cenchrus pennisetiformis*), *Tantia* (*Daeetyloctemimum sindicum*), *Shinio* (*Crotalaria burhia*), *Santo* (*Trianthema portulacastrum*), *Kanti* (*Tribulius lanugionosus*), *Bar Bordi* (*Ziziphus mauritiana*), *Kandero* (*Maytenus emarginatus*) and *Bordi* (*Ziziphus nummularia*) etc., especially during pre-seeding period has adversely affected seed production and regeneration of such palatable grasses and shrubs. Poor regeneration of palatable grasses resulted in their gradual replacement by non-palatable grasses and shrubs like *Bihani* (*Tephrosia purpurea and Tephrosia wallichii*), *Bui* (*Aerva tomentosa*), *Bekario* (*Indigofera cordifolia*), *Kheemp* (*Leptadenia pyrotechnica*), *Aak* (*Calotropis procera*) and *Ker* (*Capparis decidua*) etc. Presently major parts of Orans and Gauchars have become totally bereft of palatable grasses and shrubs resulting in heavy erosion of their capacity to support and sustain livestock population on sustained basis. Due to erosion in the capacity to provide economic benefits in the form of fodder, fuel wood and small timber etc., Oran and Gauchar, which was once a mainstay of pastoral-based rural economy, have lost their relevance in the lives of rural population to a great extent.

The average size of land holding was 17.8 ha in 1957 and has decreased to 6.0 ha at present and is likely to further decrease to 4.0 ha by 2020. Due to decrease in landholding size the traditional practice of farming is disappearing at an alarming rate and cultivation of marginal land is on the increase. The area under wastelands/pasturelands in western Rajasthan is expected to decrease from 26.4 per cent in 1900 to 19.7 per cent in 2020. While net sown area will increase from 39.4 per cent to 54.8 per cent indicating the bad state of affairs of pasture lands/fallow lands acting as forage resources for livestock.

Causes for Decline in Oran

Despite of all efforts by the government and other agencies, Orans have undergone decline and shrinkage. The following reasons may be attributed for the decline in Orans and Gauchar lands :

1. Due to rapid increase in rural population, number of landless residents who do not have land, opt for Oran and Gauchar land as their first choice. Due to poverty, lack of education, awareness about its importance and other factors, normally local residents do not object to such encroachments.
2. Orans and Gauchar provide fodder, fuel wood and also non-wood forest products but due to severe exploitation, net benefits from these areas reduce and that resulted in reduction of significance of these areas in the life of local population. This finally resulted in diversion of such areas for other purposes which led to decrease in Oran and Gauchar land.
3. Due to spread of education and rational thinking, fear of local deity to whom Orans have been dedicated, which once acted as an effective deterrence against the illegal encroachment, is gradually vanishing. Lack of any effective action against the early encroachers, also encourages many other law abiding residents to encroach the Oran and Gauchar lands.
4. In most of the government schemes for creation of rural community assets, there is no provision for making payment in respect of the land utilised for their construction. At the same time, in most of the villages, Oran and Gauchar land available there is the only community land that can be utilised, free of cost, for creation of such community assets. Though there is no statutory provision under which *sarpanch* or *gram sabha* can allot a plot of land within Oran or Gauchar land to

any individual or organisation, at many places it has been observed that *sarpanches* are allotting Oran and Gauchar lands to various persons and firms etc., by issuing *patta*. Such practices by elected representatives, without any authority of law, acts as a stimulus for encroachment of Oran and Gauchar lands.

5. Poverty and low average annual income per family along with poor literacy rate and lack of fear all added to rapid decrease of Oran and Gauchar lands.

Sacred Forestry

It is estimated that about 25,000 sacred groves and other sanctified ecosystems, varying in size between 0.1 ha to 500 ha, are in existence in Rajasthan. This is based on the extrapolation of a study based on the ratio of 1,027 recorded groves and number of villages of Rajasthan. Sacred groves are known under various names in Rajasthan as sacred groves (*deora, malvan, deorai, rakhat bani, oran,* etc.), sacred corridors (*deo ghats*), temple forests (*mandir van*) and sacred gardens (*baugh*) (Pandey and Singh, 1995, a and b).

Floral biodiversity in sacred groves is very rich. Sacred groves not only yield several non-timber forest products (NTFPs), they also harbour multiple-use livelihood goods. Resources that are traditionally obtained from trees and plants located in sacred groves include fodder, fruits, dry fallen wood, seeds and ethno-medicine. Sacred groves are the important source of water for traditional irrigation systems in Aravallis. These areas also provide habitat, water and nest-sites for wildlife and birds. Several species of honey bees nest in large trees in sacred groves (Jha *et al.*, 1998)

People do not harm sacred groves mainly because of socioreligious traditions and fear of the unknown, believing that those who cut or use an axe in a sacred grove may be harmed by the presiding deity. Sacred groves currently face various threats like submergences, clear felling, mining, quarrying, encroachment and other depletive factors. For example, a part of Ubeshwarji sacred grove in Udaipur (Rajasthan) was destroyed by submergence because of a construction of an anicut across the stream flowing through the grove. Taneshwarji Mahadev sacred grove in Udaipur (Rajasthan) is threatened by mining and stone quarrying.

Policy Issue related to Orans

Orans are unique examples of gene pool conservation based on the socio-cultural value system of Rajasthan. However, traditional biodiversity conservation methodology has not much attracted the attention of scientists, foresters and policymakers; therefore, it is an urgent need that all the existing Orans in

Rajasthan, should be systematically surveyed and demarcated by the State government. Orans are not classified and defined in the revenue records; they simply fall under 'cultivable wasteland' category. Therefore Orans should have a proper status and legislation and no allotment should be made for other purposes from the Oran land. Due to state control over Oran land, *panchayat* and local village-level committees are not making proper management and conservation of Orans. As a result, vegetation in Oran and Gauchar in Rajasthan are seriously deteriorating. The state since 1950 has followed progressive policy *vis-à-vis* cultivable lands as groves and Gaucher or pastureland all are included in the same category i.e., 'cultivable wasteland'. The forced conversion of such areas into 'revenue' allotment and regularisation of encroachment on revenue land or *siwai chak* precedence of private agriculture etc., have all been done without following provision of Forest (Conservation) Act, 1980. The communities no longer see deep sense of grievance of being robbed of a traditional entitlement. Thus, faced with severely depleted grazing lands and water sources, the people have now taken to plundering the groves. Therefore there is a need to legally and constitutionally uphold the community's right to own and manage these Orans. These lands should be declared either as forestlands or grazing lands with the ownership of communities.

During the drought year, a major amount of drought relief resources is available at the disposal of the State government and Central drought relief assistance is utilised for the supply of fodder to the starving livestock population. During the drought years millions of rupees were spent in dry fodder purchase and its transportation from neighbouring states. Even after spending so much, generally it is not possible to save entire livestock population from starvation. By undertaking suitably planned, technically feasible, fodder oriented silvi-pastoral rehabilitation measures, having proper combination of high fodder yielding perennials, Orans and Gauchars available in the western Rajasthan can act as sustainable source of fodder to meet the requirement of the existing cattle population of the desert district even during the drought years.

In the past, fear of local deity to whom Orans are dedicated acted as a credible deterrence against encroachment and illegal felling from such areas but presently the above said belief and fear are fast disappearing from many parts of the region. This may further accelerate the deterioration in the condition of the Orans. Due to increased population pressure at many places in Jodhpur district, villagers have started to encroach Orans and Gauchars. Most common types of encroachments are construction of dwelling houses and gradual extension/shifting of the boundary of the adjoining agricultural fields into Orans and Gauchars. In the absence of proper boundary demarcation, it becomes very difficult to detect encroachments in time.

At present, problem of encroachment of Orans and Gauchars is not very serious in most part of the region. But if no suitable preventive and corrective measures are initiated at this stage, in near future it may lead to serious issues in several parts of the region. Possible preventive measures may include detailed survey and boundary delineation, erection of suitable marked boundary pillars and boundary demarcation by construction of boundary wall, barbed wire fencing, planting of live hedge etc. Similarly, possible corrective measures may include eviction and shifting of unauthorised occupants of these areas through special drive by concerned *gram panchayat* and district administration. As stated earlier, presently most of these lands are under the administrative control of *gram panchayat*.

Due to non-accrual of any tangible benefit to the local population in the form of fodder, fuel wood, small timber etc., in sufficient measure, there is a growing tendency of the *gram panchayat* to divert these lands for alternative uses like rehabilitation of landless residents' construction of community assets like schools, hospitals etc. These areas not being declared as "forests" such diversion does not attract provisions of Forest (Conservation) Act, 1980. Encroachment of Orans and Gauchars by influential local residents is also common in many parts of the district. Absence of physical boundary demarcation helps in easy encroachment of these areas. Cases of the legal as well as illegal diversion of Orans and Gauchars for non-grazing purpose are fast increasing in many parts of the district. It is necessary for the conservation and protection of these lands that there must be a planned way of some special programme and that must be made with both long-term and short-term objectives.

Long-Term Objectives

i. To ensure sustainability of livestock-based rural economy of the area.

ii. To build capacity of the rural population of the area to withstand frequently occurring drought with minimum hardship.

iii. To facilitate economic upliftment of the rural areas by increasing income generation from livestock farming.

iv. To control and regulate encroachment of Orans and Gauchars which are the most important common property resource (CPR) of the region.

Short-Term Objectives

i. To increase production of fuel wood and fodder from the Orans and Gauchars available in the coming years by facilitating natural regeneration of palatable grasses, shrubs and tree species in Orans and Gauchars having sufficient root/seed stock of such palatable species by regulating/controlling grazing pressure through construction of boundary wall and also by undertaking suitable planned, technically feasible, fodder oriented silvi-pastoral measures comprising suitable combination of high fodder yielding perennials and annuals in few selected degraded Orans and Gauchars that cannot be regenerated merely by regulation of biotic interference through construction of boundary wall.

ii. To facilitate adoption of sustainable rangeland management practices like rotational grazing, deferred grazing, deferred rotational grazing etc., by dividing bigger Orans and Gauchars into small compartment through construction of boundary wall.

iii. To survey and delineate boundary of Orans and Gauchars available in the Jodhpur district, by erection of suitable marked stone boundary pillars to facilitate easy and quick detection of encroachment and to dissuade the owners of the adjoing agricultural fields from gradual shifting of boundary line into Orans and Gauchars.

iv. To create sustainable source of high quality seeds of high fodder yielding grasses, shrubs and trees for future rehabilitation works.

Keeping in view the special social and legal status, unique edaphic and vegetation properties of Orans and Gauchars, a separate 'Orans and Gauchars land policy' may be formulated by State government to ensure conservation, protection and rehabilitation of Orans and Gauchars in most cost-effective manner through active cooperation and involvement of local people, especially women, to meet their requirement of fuel wood, fodder and small timber etc., on sustained basis.

Existing statutory rules and regulations governing protection and management of Orans and Gauchars may be critically reviewed and if required, these rules and regulations may be suitably amended to make them more effective and stringent. Acts and rules governing removal of unauthorised occupants from the Orans and Gauchars may be made more stringent and effective.

On the basis of the information collected during the detailed survey of the Orans and Gauchars as outlined above, similar to the 'register of reserved forest, protected forests and un-classed forests' as maintained by various State Forest Departments in respect of the forest areas under their management, a separate register of Orans and Gauchars available in each revenue *tehsil* may be prepared and regularly updated by officer incharge of the revenue *tehsils*. In such registers, separate pages may be allotted to each Orans and Gauchars located within the respective *tehsil* and brief description of their past history as well as rehabilitation measures and diversion, if any, of its portion shall be maintained and be regularly updated from time to time.

Recommendations

Plantation programmes and other measures being undertaken for the rehabilitation of Orans and Gauchars by State Forest Department and other departments/organisations may be appropriately modified to meet specific requirement of these areas as rangeland. While planning any rehabilitation measure for Orans and Gauchars, following points may be kept in mind:

a. For undertaking rehabilitation work for Oran and Gauchar lands, local people may be taken into full confidence. The local people may be actively involved in the planning as well as execution of the work taken for rehabilitation of the Orans and Gauchars lands. The local people should be made to understand that the rehabilitation work is to be undertaken for their benefit and the implementing agency does not have any intention to grab the land.

b. Before undertaking any rehabilitation work, detailed analysis of the demand and supply position in respect of fuel wood, fodder, small timber and other non-wood forest products which can be met from the Oran and Gauchar lands as well as expectations of the local people from these, may be carried out in close cooperation and participation of the local population so that rehabilitation plan for the Orans and Gauchars may be designed to fill the identified gaps in the supply of fuel wood, fodder, small timber etc., in order of priority as decided in consultation with the local people.

c. Type of species to be planted for the rehabilitation of the Oran and Gauchar lands may be selected carefully. The species may be selected keeping in view their potential to bridge gap in the demand and supply of fuel wood, fodder, small timber and other forest products in the order of priority as decided in consultation with the local people.

d. Detailed analysis of the optimum requirement of various inputs like irrigation, farm yard manure, fencing etc., which can ensure optimum growth and survival of the species planted for the rehabilitation work, may be carried out for various species likely to be planted under different site conditions for the rehabilitation of Oran and Gauchar lands. All future proposals for the rehabilitation of the Oran and Gauchar lands may be prepared on the basis of the optimum requirement of various inputs as identified above.

e. The present practice of overreliance on hardy exotic species of the *acacias* that can survive with minimum input and care may be discontinued. In place of the exotic acacias, local species that can yield higher output of fodder, fuel wood, small timber, timber etc., and can allow palatable grasses and legume species to grow under them and can also withstand the harsh climatic conditions of the region may be preferred. Some of such local species that may be gien more preference in such rehabilitating works are *khejri* (*Prosopis cenercria*), *rohida* (*Tecomella undulata*), *kankera* (*Maytenus emarginatus*), *neem* (*Azadirachta indica*), *kumat* (*Acacia senegal*), *ardu* (*Ailanthus excelsa*), *mopane* (*Colophospermum mopane*), *anjan* (*Hardwickia binata*) and *desi babool* (*Acacia nilitica*) etc.

f. In order to prevent the recurrence of problems caused by the exotic species like *Prosopis juliflora*, no new exotic species may be introduced in the region without detailed analysis of their impact on the native flora in general and desert ecosystem in particular.

g. As far as possible, plantation may not be raised in the portion of the Oran and Gauchar constituting catchment area of the village pond, locally called *agor*. However, due to certain unavoidable reasons, plantations are required to be raised in the *agor*, it shall be raised without extensive soil working that may reduce run-off rate of the rain water resulting in the poor recharging of the village pond. Species like *Prosopis juliflora* and any other species whose litter fall may cause deterioration of the water quality of the village pond may not be planted in the *agor*.

h. Suitable contingency plan to cope with the deficit rainfall during the drought years may be a compulsory constituent component of each project/scheme formulated for the rehabilitation of the Oran and Gauchar lands. Such contingency plan may contain provisions for the additional watering of the plants during the drought years to avoid large scale mortality during such drought years.

i. Adequate provision may be made for the protection of the plants raised under rehabilitation programmes, from the termite attack.

Conclusion

In view of above facts and conditions, it is an urgent requirement not only for the government or non-government agencies but also of common people to preserve their rich Oran and Gauchar lands. Rajasthan is known for its rich cultural and religious beliefs and if these Oran and Gauchar lands are not protected then the condition of the people who often depend on these lands during famine, will be dreadful. Protection of these lands means simply protection for future generation.

References

Chandrakanth, M.G. and Jeff Romm (1991). "Sacred Forests, Secular Forest Policies and People's Actions", *Natural Resources Journal* 31: 741-756.

Gupta, J.P. (2003). "Management of Community Grazing Lands for Security and Improved Living", Proceeding of workshop on *Development of Suitable Strategies for Rehabilitation of Oran and Gauchar in Rajasthan* at AFRI, Jodhpur, pp. 9-19.

Jha, M., H. Chatterjee Vardhan, K.S. Kumar and A.R.K. Sastry (1998). "Status of Orans (Sacred Groves) in Peepasar and Khejarli Villages in Rajasthan", in P.S. Ramakrishnan, K.G. Saxena and U.M. Chandrashekara (eds.), *Conserving the Sacred for Biodiversity Management*. New Delhi/Calcutta: Oxford and IBH. pp. 263-275.

Kumar, P., N. Bala, G. Singh and S.P. Chaukyal (2003). "Status of Common Access Resources (CAR's) and Socio-economic Conditions in Some Villages of Rajasthan and Gujarat", Proceeding of workshop on *Development of Suitable Strategies for Rehabilitation of Oran and Gauchar in Rajasthan at AFRI*, Jodhpur. pp. 97-102.

Mertia, Ramsingh and Rajendra Prasad (2003). "Oran Gauchar Land in Jaisalmer: Present condition and need for proper management", in *Maru Krishi Chayanika* (Oran – Gauchar Species issue) Published by CAZRI (in Hindi). pp. 62-66.

Narayan, Pratap, J.R. Sharma, Suresh Kumar and Charan Madhu Bala (2003). "– Oran and Gauchar – Sanderb and Paripaksh", in *Maru Krishi Chayanika* (Oran – Gauchar Species issue). Published by CAZRI (in Hindi). pp. 1-8.

Pandey, Deep N. and Samar Singh (1995a). "Aravalli Ke Deovan", *Rajasthan Patrika*, 21 May.

———. (1995b). "Traditions of Sacred Groves in Aravallis", *Wastelands News*, (Hindi), April-June, pp. 3-6.

17

Forest Dependence and Depletion

Socioeconomic Analysis of Forest Protection Committee Members in the Bankura District of West Bengal

SOUMYENDRA KISHORE DATTA and KRISHANU SARKAR

Introductory View

There is no doubt that in recent years prudent management of local common pool resources/CPRs has assumed a great importance from the viewpoint of providing means of livelihood to a great number of rural poor people. Wetland, fishery, forest resources, grazing land, groundwater etc., constitute important CPRs capable of providing wherewithals of life to a vast section of people. Among these most important appears to be the forestland for its diverse types of timber and non-timber benefits including environmental benefit. In India forest resources constitute the second largest land use after agriculture with coverage of 641.13 sq km or 22 per cent of total land mass. Apart from providing diverse benefits to the economy in general, forest resources provide part of the subsistence and cash livelihoods to about 275 million rural poor in the country. The gradual depletion of forest cover and increasing dependence of rural folk has given rise to a sort of conflict of interest among diverse user groups and associated management problems. The primary stakeholders are the forest communities or people living in and on the forest fringe areas earning substantial part of livelihood in the form of fuel wood, fodder, fruits, flowers, fencing materials etc. The secondary stakeholders are state/district Forest Department, NGOs, plantation companies etc. The traditional rights and interests of the forest communities, and their livelihood and primary stakeholders' concerns were hardly addressed to in the forest management policy before and immediately after independence. The early '90s witnessed the introduction of Forest Conservation Act, 1980 with a shift towards conservation. As a follow up, this was vindicated by introducing a number of experimental steps by West Bengal and some other states through allocation of specific areas of forest along with limited management responsibilities to forest communities in return for a share of timber revenue and access to non-timber forest products (NTFPs).

Since the enunciation of Forest Protection Act, the traditional community access and control has been substantially wrested by the Forest Department in West Bengal and in some other states with only specific duties and functions reserved for the forest protection committee (FPC) formed by the villagers. But this has hardly reduced the poor people's dependence for collection of NTFPs that provide a substantial part of their livelihood in regions that are less agriculturally intensified. Apart from this, the enjoyment of CPR resources by non-poor households is not at all negligible. However since the joint forest management (JFM) institution is highly tilted in favour of dominance of forest department control over planning, management, investment, harvesting and marketing, the forest dwellers often fail to identify their customary rights with forest resources and tap its sustainable productivity and livelihood potential. In the absence of strong forest rights and management responsibilities and lack of monitoring by state agency, communities often stray away from conservation goals and their activities fail to protect the forest from being degraded. It is estimated that about 41 per cent of India's forest has been degraded in the past decades.

In developing countries degradation of CPRs like forests has often been associated with high rate attached by the poor in discounting future flow of benefits, improper property rights, economic insecurity, regulation failure and relative ability of local institutions that shape the degree of collective action. In regions where variations in local institutions do not exist, degree of difference in people's participation in conjunction with differential in socioeconomic factors determines the common property outcomes. Insecure property right and associated risk of eviction, poverty and poor access to insurance and credit market are often viewed as underlying factors responsible for degradation of CPRs. Again absence of motivation/monitoring often fails to enforce active participation thereby leading to degradation of the resource. Apart from this lack of alternative income sources, rising population size, insufficient local employment opportunities, lack of social capital like mutual trust and amity as well as high dependence on forest as a safety need during difficult economic times lead to unsustainable management of the resource. Sometimes long-term interest of the community is sacrificed due to short-run economic pursuits. The problem becomes grave when despite degradation and shrinking availability and restricted access, resource dependence of the poor does not diminish. In this backdrop the aspect of insecurity in property rights, dependence on forest, dynamics of socioeconomic features, relative availability of alternative job and income availabilities and tradeoff between short- and long-term interest of the community merit an integrated study in a highly CPR dependent region exposed to the risk of degradation. The problem gets added importance when studied in respect of a CPR region inhabited mostly by backward community who are usually supposed to be highly dependent on nature for their survival.

Location of Study Region

The district of Bankura is situated on the western side of the district of West Bengal bordering on the Jharkhand state. According to 2001 census, about 92 per cent of the people in the district live in rural areas. Tribal people constitute about 10.33 per cent of the district population. Thus total forest area under legal status in Bankura stands as 1482 sq km (about 21.5 per cent of the district) which is below the national average of 33 per cent forest coverage. This forest area is broadly divided into two divisions, Bankura South (747 sq km) and Bankura North (735 sq km). Recently there has emerged a smaller third division called Panchet division.

The following table reveals that in 1994 the number of FPCs as well as protected area in northern division was about 2.5 times compared to that in southern division. However this disparity almost leveled up in 2006. In terms of sex-wise as well as caste-wise FPC membership however, northern division surpassed the southern division. If we compare average number of members per FPC we find that in northern region there are about 264 members compared to only 94 in southern division while the absolute number of FPCs are greater in the latter compared to the former. This indirectly indicates that there are comparatively greater degrees of heterogeneity in terms of geographical locations, economic status, social order as well as interest of the members in the southern division and the formation of FPC had to be based keeping in mind the possible homogeneous elements in an otherwise heterogeneous scenario. Table 17.1 depicts a comparative scenario of number of FPCs, protected area, gender and caste status of members in FPCs across two forest divisions in Bankura district and the state as a whole.

Usually, comparatively economically weaker sections and indigeneous tribal people are highly dependent on forest resources and their FPC membership is about 54.5 per cent in the northern division while 55 per cent in southern division. Interestingly, however, there has been observed a tendency towards increased encroachment on forestland for agricultural purposes in southern division compared to a reversal in the northern division as is evident from Table 17.2. This indicates a tendency towards greater degradation of forest resources in the southern part of the district compared to that in the northern part.

Table 17.1

Comparative Scenario of Number of FPCs, Protected Area, Gender and Caste Status of Members in FPCs in Bankura Forest Region and West Bengal

Division			Bankura (N)	Bankura (S)	West Bengal
Total no. of FPCs		Mar'94	609	256	
		Mar'06	547	593	4079
Area protected (ha)		Mar'94	60038	21941	
		Mar'06	43613.19	43191.16	542057.217
No. of members	Male	Mar'94			
		Mar'06	138990	51888	519551
	Female	Mar'94			
		Mar'06	5400	3946	43077
	SC	Mar'94			
		Mar'06	57304	13118	177812
	ST	Mar'94			
		Mar'06	21494	17906	134594
	Others	Mar'94			
		Mar'06	65592	24810	250222
	Total	Mar'94	60027	17689	
		Mar'06	144390	55834	562628

Table 17.2

Encroached Forest Area in Bankura Forest Divisions

Forest division	Area Under Encroachment (ha)	
	December 1987	March 2006
Bankura (N)	1516.00	1406.68
Bankura (S)	2112.00	2300.39
Panchet soil conservation	N.A	1331.51
Bankura	N.A	5038.58

Based on this comparative scenario of northern and southern divisions, it seems imperative to focus on the southern part because of its greater forest cover, supposedly higher degree of heterogeneity, increasing tendency of forest area encroachment among dwellers etc.

Bankura south division is composed of 12 forest ranges, each forest range having a number of forest beat offices for administrative convenience. Each beat again serves a number of villages. In this paper we consider a part of the Ranibundh and Punshya forest beat in the southern forest division of Bankura district for our study. In Ranibundh beat, 7 villages have been selected and in Punshya beat 3 villages have been selected. From each village 50 per cent of the FPC members have been selected based on convenience sampling.

Forest Dependence for Self-Consumption

The beneficiaries of JFM in the study region are mostly poor households belonging to scheduled caste (SC) and scheduled tribe (ST) class and are dominated by landless labourers, marginal and small farmers and agricultural workers. The exigency of meeting their immediate basic survival needs and impaired capability, often compel them to resort to overexploitation of forest resources despite their awareness of the importance of conserving the forest resources. Situation like 'tragedy of commons' sets in when there is poor monitoring of illegal access to forest resources and connivance of Forest Department officials as well as FPC members who themselves are offenders.

The following table provides a village-wise view of the level of forest dependence (represented by self-consumption values) in Ranibundh beat as explained by several explanatory variables. Extent of self-dependence on forest products in each village is assessed by regressing the individual self-dependence values on local employment days (LED), status(S), respective family size (FS), income other than agriculture (OI), number of items collected from forestland (NI) and livestock (LS).

In case of Ranibundh beat overall regression is significant in all cases as evident from the following R^2 values in Table 17.3. Major significant variables appear to be OI, FS and NI and their signs also are in the expected direction. As family size increases, villagers also increasingly depend on CPR for livelihood. With increase in income from other sources there is an expected decline in the degree of forest dependence on the part of the villagers. And in case the forest happens to be rich in supplying an increased variety of species, its exploitation for survival needs is increased leading to a higher level of dependence.

In case of Punshya beat also, as shown in Table 17.4, self-consumption regression equations happen to be significant for all the considered villages with FS, NI emerging as major significant explanatory variables with their signs in the expected direction.

Table 17.3

Least Square Parameter Estimates of Variables Influencing Self-Consumption in Ranibundh Beat

Variable	Village						
	Katiam	*Buriam*	*Keliapathar*	*Bhurkura*	*Mahadevisinan-Barudi*	*Barapocha*	*Jamdaha*
LED	0.88 (0.44)	2.23* (0.07)	-0.05 (0.92)	0.16 (0.89)	2.84 (0.51)	-1.54 (0.32)	8.47* (0.01)
S	-126.08 (0.66)	-451.14* (0.09)	140.6 (0.38)	55.02 (0.76)	-39.95 (0.89)	182.01 (0.44)	–
FS	489.16* (0.00)	431.12* (0.00)	411.13* (0.00)	265.35* (0.00)	457.02* (0.03)	500.49* (0.00)	504.98* (0.00)
OI	-0.0065* (0.05)	-0.0039* (0.06)	-0.0043 (0.31)	-0.0024 (0.67)	-0.0089 (0.41)	-0.0058* (0.00)	-0.04* (0.04)
NI	251.59 (0.12)	587.94* (0.00)	670.54* (0.00)	570.71* (0.00)	348.10 (0.28)	674.74* (0.00)	1643.70* (0.00)
LS	-3.7 (0.86)	5.81 0.71)	-18.91 (0.31)	13.32 (0.36)	-7.81 (0.79)	-60.96* (0.02)	51.65 (0.26)
R^2	0.59	0.62	0.69	0.61	0.53	0.81	.80
F	14.86* (0.00)	11.26* (0.00)	19.19* (0.00)	7.47* (0.00)	2.52* (0.07)	20.84* (0.00)	10.73* (0.00)

Note: * value of parameter estimates are significant at respective levels given in the parenthesis.

Table 17.4

Least Square Parameter Estimates of Variables Influencing Self-Consumption in Punshya Beat

Var	*Jantadumur*	*Kotra*	*Dhankura*
LED	-1.02 (0.55)	0.55 (0.78)	0.20 (0.95)
S	-231.59 (0.54)	158.95 (0.28)	-432.75 (0.56)
FS	486.77* (0.00)	442.38* (0.00)	168.90 (0.47)
O I	0.0008 (0.89)	-0.01 (0.19)	0.008 (0.52)
NI	452.11* (0.02)	218.49* (0.03)	1034.83* (0.00)
LS	-12.80 (0.55)	6.72 (0.48)	11.29 (0.64)
R^2	0.61	0.61	0.59
F	8.74* (0.00)	9.47* (0.00)	6.03* (0.00)

Note: * value of parameter estimates are significant at respective levels given in the parenthesis.

However despite the emergence of JFM system and villagers' forest dependence, because of lack of coordination between Forest Department and FPC, politicisation of FPC activities, widespread illiteracy and negation of customary rights to access the forest resources, villagers often do not feel motivated to protect the forest with a feeling of alienation from their traditional property rights. As a result there has been observed a tendency towards breaking the rules of access, illegal felling, and collection of different species at a rate beyond their survival capacity and reducing the forest base to prepare land for quick return yielding *babui* grass cultivation. The rate of exploitation of different trees, plants and shrubs had been often much higher than the flora's regeneration. The destructive impact of this rate of consumption is especially discernable in the case of less-frequent species. Again introduction of high yielding *swarna* paddy had an expansionary effect on peoples' attitude to establish tenurial rights over more cultivable land with relative neglect of perennial herbs and other species of plants. The consequential increase in livestock and their grazing in the areas adjoining forestlands had a deleterious effect on the growth of young shoots and tender herbs of different species. Regeneration of various slow growing species was possibly curtailed as the green aerial parts were quickly devoured and growth stunted by the trampling of hooves. Apart from this, occasional manmade fire-burning of specific areas of forestland to make room for clear collection of *mahua* and *kendu* seeds, leaves and fruits had damaging impact on the growth of some tender species of plants. The different species that contribute in many ways to the earning of livelihood and medical benefits to the indigeneous people in the study area is depicted in Table 17.5.

Table 17.5

Different Plant Species and their Uses

Species	*Uses*
1. *Sal*	The leaves are used to make plates; it is also used as construction material for their house, furniture, fence, posts etc.
2. *Kendu*	The leaves are used in combination with tobacco for making smoking items and the plant is used as firewood also.
3. Bamboo	From selling bamboo people can earn money directly; it is also used to make items like basket, handicraft goods and broom.
4. *Arjun*	It is used to cultivate *tasar* that is sold at a high price in the market; the liquid extract from the tree is use as *gala*, the bark is used as a medicinal herb.
5. Eucalyptus	It is used for firewood collection (it is high yielding and high-growth tree); the stem is used for constructing houses.

contd...

...contd...

	Species	*Uses*
6.	*Mahua*	Berries are eaten, made into local alcoholic beverage, people sell alcohol and earn money, from its berries oil can be extracted (collected) which is used in cooking, the dry refuse of fruits is used as fodder.
7.	*Kendh*	Berries are eaten, from selling fruits people can earn money, used as a firewood also.
8.	*Bhurru*	Edible sweet fruit.
9.	Mango	Edible sweet fruit.
10.	*Khejur*	Edible sweet fruit and liquid extract from the plant is used as soft drink.
11.	*Susuni sak*	Edible green plant leaf.
12.	*Shojne/sak*	Edible green vegetable/plant leaf.
13.	*Neem*	It is generally used in construction, a twig used as a toothbrush; the leaves are used as food.
14.	*Bel*	Edible sweet fruit, it is used as a medicine herb also.
15.	*Kalmegh*	Used as a medicinal herb.
16.	*Bhaluk sakti*	Used as a medicinal herb.
17.	*Rahara*	Fruits of this plant are used as medicinal herbs.
18.	*Bara gutur*	Used as a medicinal herb.
19.	*Haritaki*	Used as a medicinal herb.
20.	*Rerimakha*	Used as a medicinal herb.
21.	*Loadh Chal*	The bark of this tree is used as medicinal herb.
22.	*Harrura Chal*	The bark of this tree is used as medicinal herb.
23.	Aloe vera (100 types of species; according to villagers)	Used as a medicinal herb. Especially it helps increase the vitality of life. (locally it is known as *ghritokumari*)
24.	*Satmul*	Medicinal tea is made from the root of this plant.
25.	*Anata mul*	Medicinal tea is made from the root of this plant.

Source: Field survey.

Decline of Species and Region-Specific Forest Degradation Factors

The fact that species have declined in most of the sample villages together with reported forest degradation, is revealed from the following Tables 17.6 and 17.7. During lean periods marked by privation and physical stress, villagers are left with no other alternative than to fall back upon forest more out of compulsion than choice. Apart from illicit felling and cutting of trees/branches, poor households continue to access forests for collecting various NTFPs together with grazing their cattle. The FPC members themselves often ignore mutual illegal collection from forest out of sheer necessity of survival. This sets forth the problem of 'tragedy of the commons' with above poverty line (APL) class people also somewhere preempting NTFP collection in order to be served first on first-come basis. Apart from this, females are very thinly represented in the elected boards of most of the FPCs. They also hardly participate in the FPC meetings and are rather weakly motivated to protect the forest resources. Most of the NTFP collection activities are done by the females while spending a lot of time after it. The following tables reveal the apparent connection between perceived threat of eviction, forest degradation and decline in species.

Table 17.6

Percentage of Respondents Perceiving Species Decline, Forest Degradation and Sensing Eviction in Ranibundh Beat

Village	*Species Declined (%)*		*Village Forest Degradation (%)*		*Perceived Threat of Eviction (%)*	
	Yes	*No*	*High*	*Low*	*High*	*Low*
Katiam	91.04	8.95	85.07	14.93	80.60	19.40
Buriam	95.74	4.26	80.85	19.15	68.08	31.92
Keliapathar	53.45	46.55	48.28	51.72	43.10	56.90
Bhurkura	94.28	5.72	80	20	80	20
Barapocha	83	17	81	19	75	25
Mahadevisinan-Barudi	85	15	80	20	80	20
Jamdaha	0	100	0	100	31	69

Source: Field survey.

Table 17.7

Percentage of Respondents Perceiving Species Decline, Forest Degradation and Sensing Eviction

Village	*Species Declined*		*Village Forest Degradation (%)*		*Perceived Threat of Eviction (%)*	
	Yes	*No*	*High*	*Low*	*High*	*Low*
Jantadumur	40 (100)	0	40 (100)	0	40 (100)	0
Kotra	44 (100)	0	44 (100)	0	44 (100)	0
Dhankura	30 (93.75)	2 (6.25)	30 (93.75)	2 (6.25)	30 (93.75)	2 (6.25)

Source: Field survey.

Major percentage of people in all the villages except Jamdaha reported a decline in the species. This percentage is observed to be highly in accordance with that of perceived high degradation of forest on the part of the respondents. Again excepting Keliapathar and Jamdaha, majority of the villagers in other areas are observed to be reeling under the threat of eviction and hence insecurity of their forest right. This is likely to be accentuating the process of degradation. Apart from that there are certain village-specific factors that accentuate the process of depletion and degradation and these emerge mostly due to lack of villagers' commitment and sensibility to consider forest as their own property and traditional means of survival. For instance, in Katium, illegal plundering and cutting of tress has been rising day-by-day according to the villagers, leading to gradual depletion of the forest. The Sabar people here are generally treated with neglect in FPC meetings and accordingly they find little incentive to put labour in forest protection and preservation. In case of Bhurkura, Forest Department often undertakes cutting *sal* trees and FPC people burn their shoot in order to plant quick return yielding eucalyptus trees which also reduce the availability of *sal* leaves. The spread of eucalyptus tree has thinned the forest of other species since this plant impedes the growth of other trees. Non-availability of *sal* leaves here drive poor villagers in this locality to collect *sal* leaves/other twigs, often illegally from nearby forests under control of different FPCs. In Mahabarudi village some forest people undertake quick felling of *kendu* plant at the stage of their less than moderate growth in order to produce lasting cot support legs. Again in case of growth of other plant species, *kendu* plant leaves suffer from lack of adequate sunlight leading to its gradual non-availability. Since *kendu* leaves are economically productive, villagers often resort to clear cutting of other plants in order to support *kendu* leaf generation. Lack of regular monitoring and improper imposition of punitive measures have led FPC members in the aforesaid region to take to unrestricted degradation of forest without coordinating much with Forest Department activities. Again, with the spread of the news of a possible inroad of sanctuary here, villagers have started to panic with a gradual loss of sense of belonging to the traditional forest area. Lack of adequate agricultural credit also often deters the forest people from generating subsistence level of crop in their land. The consequent alternative for village people is to heavily deplete the forest often out of desperate attempts by flouting and transgressing the forest protection rules, even at the risk of being caught. The villages in Ranibundh had been very rich in terms of various types of medicinal plants like *satmul, anata mul, ghritokumari,* kalmegh, etc. In recent times, marketing channel of these species has been greatly improved here through the inroad of two agencies who are engaged in collecting them for commercial uses. This has enhanced the rate of depletion of these species which are cut by the village people at their earlier stages in order to reap quick profits by selling them to these agents almost at their doorsteps. As a result there is an adverse stunting impact on the regeneration of these medicinal plants and loss of rich biodiversity.

Spread of commercial cultivation of *babui* grass by clearing forest as well as illicit rapport between forest contractors and FD staff often lead to withering away of huge chunks of high value forest resources endowed with a multiple variety of species. *Babui* grass is a quick growing species, used for making ropes and has a good market. Therefore from economic point of view, in order to derive short-run gains villagers often resort to illegal cultivation of this grass by encroaching on forestland which have a very deleterious effect on the long-run productivity of soil as the land can never be used for growing crops or any other plant species. Due to illegal colluding connection between some forest dwellers, forest contractors and connivance of Forest Department staff as well as lack of proper care and maintenance, the number of species and availability of forest produce declined jeopardising further the prospect of a rural forest-based livelihood with a high-perceived degradation and threat of eviction. Unchecked depletion of forest resources for deriving short-run benefits is thus associated with reduced species diversity as well as reduced efficacy of forest as a carbon sink leading to long-term diverse uncertain environmental consequences. The problem of tradeoff between short-run economic gains and long-run environmental cost is discernible from the following dynamic optimisation model.

Tradeoff between Short-run and Long-run Benefit/Cost Effects of Forest Extraction

In order to describe the nexus between short-term benefit and long-term effect, the following model seems to suit the situation of forest depletion in the study region. Short-run benefit is assumed to be influenced by the amount of collection from and extraction of forest resource (f), the stock of different type of species (s) inclusive of animals and plants as well as the level of ambient environment ($\underline{c}$- c). Here $\underline{c}$ is the environmental carrying capacity while c is the stock of carbon. So the forest benefit function can be written as

$$B = \int_0^\infty u[f_t, s_t, (\underline{c} - c_t)]e^{-rt}dt, \quad \text{Let } \underline{c} - c_t = c'_t$$

Here change in carrying capacity net of carbon stock is a continuous inverse function of extracted forest. i.e.,

$$dc'_t/dt = g(f_t) \quad \text{where } g_f < 0$$

Change of species stock is assumed to be a function of extracted forest where increased deforestation (forest extraction) has the impact of reducing the stock of species (biodiversity). Hence species stock growth is negatively related with extracted forest. Following this we have:

$$ds_t/dt = h(f_t) \quad \text{where } h_f < 0$$

Now our aim is to establish the relationship between short-term benefit which is coming from extracted forest and the long-term effect based on the impact of change in stock of species and carbon dioxide. This follows from maximising benefit function subject to the constraints.

$$\text{Max} \quad B = \int_0^\infty u(f_t, s_t, \underline{c} - c_t)e^{-rt} dt$$

Subject to:

$$ds_t/dt = h(f_t) \quad \text{where } h_f < 0$$

$$dc'_t/dt = g(f_t) \quad \text{where } g_f < 0$$

The corresponding current value Hamiltonian stands as :

$$H_c = U[f_t, s_t, (\underline{c} - c_t)] + \lambda_{1t} h(f_t) + \lambda_{2t} g(f_t)$$

By optimal control condition :

$$\delta H_c / \delta f_t = U_f + \lambda_{1t} h_f + \lambda_{2t} g_f = 0$$

$$\text{or} \quad U_f = -\lambda_{1t} h_f - \lambda_{2t} g_f \qquad \text{...(A)}$$

$$\text{and} \quad d\lambda_{1t} / dt = r\lambda_{1t} - \delta H_c / \delta s_t \qquad \text{...(B)}$$

$$d\lambda_{2t} / dt = r\lambda_{2t} - \delta H_c / \delta c'_t \qquad \text{...(C)}$$

Where U_f represents marginal benefits from extracted forest . It is positive.

$\lambda_{1t} h_f$ represent shadow value of marginal change or loss in species due to forest cutting or extracted forest. Since $h_f < 0$, the 1st term (with sign) on RHS in (A) is positive. $\lambda_{2t} g_f$ represent shadow value of marginal change or worsening of ambient environment due to increase in carbon emission resulting from extracted forest. Since $g_f < 0$, the 2nd term (with sign) on RHS in (A) is also positive.

At steady state optimal depletion condition both $d\lambda_{1t} / dt$ and $d\lambda_{2t} / dt$ are equal to value zero.

Therefore from (B) $r\lambda_{1t} = \delta H_c / \delta s_t$ as well as from (C) $r\lambda_{2t} = \delta H_c / \delta c'_t$

or we get $\lambda_{1t} = (\delta H_c / \delta s_t) / r$ and $\lambda_{1t} = (\delta H_c / \delta_{st}) / r$

So, putting these values in (A) we get:

$$U_f = -[(\delta H_c / \delta s_t) / r].h_f - [(\delta H_c / \delta c'_t) / r].g_f$$

The implication is that at steady-state optimality situation, discounted marginal loss of utility due to species reduction resulting from deforestation together with discounted marginal loss of utility due to damage to ambient environment due to deforestation must be equal to the gain in marginal utility due to the increased deforestation. It further implies that at equilibrium due to gain in immediate benefits from deforestation there is a strong cost aspect in terms of reduced species stock along with reduced efficacy of forest resource as a carbon sink in the longer run. The negative environmental implications of reduced species stock and reduced quality of ambient environment are really alarming from the point of view of sustenance of the resource community.

The Determinants of Perceived Forest Degradation

To assess the impact of sense of fear of eviction any time by the higher authority and some other relevant variables on the attitude towards status of degradation of the forest resources, a case study was carried out in 2008 in the aforesaid villages in southern forest division of Bankura district in the state of West Bengal. In each village 50 per cent of the village forest dwellers were selected as the respective sample size. Degradation or no degradation is considered as dependent dummy, linearly related with explanatory variables in the form

$$L_n[P/(1-P)] = \alpha + \beta_1 SV + \beta_2 FF + \beta_3 FS + \beta_4 OI + \beta_5 PL + \beta_6 DE + \beta_7 SD + \beta_8 MT$$

Where P is the probability of degradation of natural resource, sale value = SV, fear of eviction in future = FF(dummy, with perceived threat of eviction assigned value 1 and its absence assigned value 0), family size = FS, other income = OI, poverty level = PL, local days employed = LDE, level of forest dependence = FD (dummy, with annual income from forest greater than Rs.8000/- being imputed value 1 and income less than this imputed value 0), mutual trust = MT (dummy, with perceived existence of mutual trust assigned value 1 and its absence assigned value 0). One village is not considered in this analysis as no degradation was recorded there. Binary-Logit regression gives the following results in Table 17.8.

It is observed that fear of eviction (FF) appears to be the most significant variable responsible for the attitude towards forest degradation in all the villages except Katium. Poverty and mutual trust though significant in some cases, vary in sign. The implication is that poverty level may not always be linked directly with forest degradation. Similarly as absence of mutual trust may lead to predation of the resource without any mood of conservation, its presence often expedites it when everybody can bank on others' mutual sympathy for similar depletion. Count R^2 values are rather high in all cases implying that considered variables amply influence perceived attitude towards degradation. Jamdaha village is not considered in the above regression as there is no reported forest degradation. Again, since in all the sample villages in Punshya beat degradation was reported by almost 100 per cent of the villagers, aforesaid regression could not be carried out.

Table 17.8

Village–wise Results of Parameter Estimates of the Logit Regression in Ranibundh Beat

Variable	*Village*					
	Katiam	*Buriam*	*Keliapathar*	*Bhurkura*	*Mahadevisinan-Barudi*	*Barapocha*
SV	7.31E-05 (0.6425)	5.29E-05 (0.5821)	-0.00076* (0.0147)	0.000895* (0.0633)	-0.000229 (0.5901)	9.99E-05 (0.7991)
FF (dummy)	1.249227 (0.2400)	1.343494* (0.0140)	2.907517* (0.0003)	6.526283* (0.0000)	7.434957* (0.0469)	2.829670* (0.0339)
FS	-0.197113 (0.3271)	0.164863 (0.2870)	0.594966* (0.0779)	0.346688 (0.6612)	2.878819 (0.1068)	-0.433390 (0.2522)
OI	-2.43E-05* (0.0032)	2.14E-06 (0.7466)	2.21E-06 (0.8939)	4.05E-05 (0.4881)	-0.00104* (0.0310)	1.38E-05 (0.4579)
PL	0.18083* (0.0010)	0.04632 (0.1832)	-0.12670* (0.0168)	-0.34763* (0.0246)	-0.63239 (0.2267)	-0.11386 (0.3442)
LED	-0.01560* (0.0174)	0.004172 (0.2420)	-0.005384 (0.1211)	0.013653 (0.4332)	-0.26903* (0.0041)	0.014663 (0.3345)
FD (dummy)	-0.458888 (0.6679)	-1.67302* (0.0620)	-27.8303* (0.0000)	-9.836224* (0.0857)	-17.5724* (0.0080)	-0.908153 (0.7586)
MT (dummy)	-1.646296 (.2834)	-0.249410 (0.7243)	1.673467* (0.0705)	0.747460 (0.6113)	-12.2142* (0.0129)	1.129541 (0.4347)
Log likelihood	-22.31674	-15.91467	-22.82487	-7.871171	-5.733786	-12.04792
Mc-fadden R^2	0.209597	0.306660	0.431765	0.550580	0.427083	0.320621
Count R^2	.8806	.8936	.7931	.9143	.8000	.8337

Note:* value of parameter estimates are significant at respective levels given in the parenthesis.

The fact that degradation has been greatly influenced by the threat of eviction from forestland is evident from the regression results. Most of the villagers are very poor, have little land of their own and feel insecure at the slightest indication of eviction from their traditional accessed land or spread of the idea of potential developmental projects through clear cutting of forest-lands. In such circumstances the available forest resources tend to be treated as open access resources for preempting the corresponding benefits while the forest is pushed to the state of depletion and degradation. Most of the villagers are very poor in terms of land ownership(both in quantitative and qualitative sense) and hence are likely to add to their plot and thus try to establish their right at the slightest indication of future eviction from their land. For categorising the villagers in terms of ownership of different quality of agricultural land and size of landholding, some basis of classification was adopted. If the productivity of land appears as 16 quintal rice per *bigha*, it is considered as low quality, land yielding production in the range of 16 to 20 quintal per *bigha* is considered as average while greater than 20 quintal per *bigha* is classified as high quality. In case of land size, ownership less than 10 *bigha* is classified as low, 10 to 16 *bigha* as average while greater than 16 *bigha* classified in high group. The data in Table 17.9 accords with the view that major percentage of the villagers in all the considered villages, having ownership of small pieces of poor quality agricultural land feel very insecure in their existing status and hardly bother about forest enhancement strategies requiring increased investment of efforts and inputs. Instead, any impending threat of eviction goads them to extract more from forestlands for guarding against their uncertain future livelihood prospects.

Table 17.9

Characteristics of Agricultural Land in the Sample Villages in Ranibundh Beat

Determinants	*Katiam (%)*			*Buriam (%)*			*Keliapathar (%)*		
	Low	*Average*	*High*	*Low*	*Average*	*High*	*Low*	*Average*	*High*
Quality of agricultural land	100	0	0	100	0	0	98.27	1.73	0
Size of agricultural landholding	89.55	7.46	2.98	87.23	10.64	2.12	98.27	1.73	0

Determinants	*Bhurkura (%)*			*Mahadevisinan-Barudi (%)*			*Barapocha (%)*		
	Low	*Average*	*High*	*Low*	*Average*	*High*	*Low*	*Average*	*High*
Quality of agricultural land	100	0	0	100	0	0	100	0	0
Size of agricultural landholding	97.14	2.86	0	95	5	0	100	0	0

Source: Field survey.

Concluding Observations and Policy Suggestions

There is international support to the observation that peoples' participation and forest conservation is greatly improved through emboldened security and enhanced forest rights and responsibities of the forest dwellers. Although about 100 million indigeneous people across the world live in and around forest areas, most of them are deprived of a secured access to such resources and their customary rights are often left at stake due to institutional non-recognition. In a changing global environmental perspective, the importance of providing a stronger forest right and secured tenure over natural resources has been perceived by many national governments. Although JFM was instituted in India in order to decentralise forest management responsibility by recognising communities' attachment and their indigeneous knowledge of conservation, various state governments have devolved only limited management responsibility and allowed restrained access to forestland with a modest usufruct rights to the forest-based communities. Many villagers view JFM as a top-down non-participatory process in which peoples' traditional rights, needs and aspirations are hardly addressed to. JFM has, in many cases, failed to effectively encompass the need for strengthened security of tenure and legal recognition of their enhanced rights to access and use the natural resources. In its absence, communities usually lack incentive to invest in long-term conservational management and their spiritual links with forest resources tend to get loosened. According to Molnar *et al.* (2004), secured tenure over natural resources and strong access rights are crucial for psychic and physical involvement of villagers in saving the resources from depletion. Many findings from Kenya, Ecuador, Indonesia, southern Africa and other places (Aihoon and Kirsten, 1994; Roth and Haase, 2000; Burkard and Ebersberger, 2002) support the importance of tenure security in influencing investment in land and checking depletion of natural resources. In certain studies (Bedoya, 1987; Hanna and Munasinghe, 1995) it is stressed that deforestation and non-sustaining agricultural practices are associated with incomplete or non-enforced property right. In this context it may be said that the FPCs also have not been endowed with adequate power commensurate with their duties and functions as specified in 1990 Government of India (GoI) resolution. Absence of the power to penalise the offenders cripples them in effectively taking harsh measures to thwart illegal forest-based activities. The FPCs feel helpless and demoralised when they come into conflict with offenders. They cannot take decisions regarding the conservation measures independent of views of FD. In order to promote active peoples' participation in forest management, state forest policies need to be oriented towards recognition of historical tenure-based rights. This should be governed by the principle of chalking out a new joint community management framework with stronger forest resource tenure and power for communities while the FD relegated to the role of indicating, identifying and guiding the communities towards conservation goals.

Although states were encouraged to acknowledge historic rights in a 2000 circular from the Ministry of Environment and Forests (MoEF), the process of implementation has been a rather slow affair. Usually the issue of tenure remains rather vague in JFM resolutions and accordingly Forest Department gets a bigger hand in its governance. The importance of assigning an unambiguous, secure and exclusive forest rights to communities for a defined area of forest has not been focused in the JFM resolutions. According to Molnar *et al.* (2004) efficient forest rights to communities should cover the following aspects:

(i) Forest rights should be clearly defined.

(ii) The duration of the right is clearly delineated.

(iii) The rights cannot be unilaterally taken away.

(iv) The boundaries are clearly specified.

(v) The notified government agency must be entrusted with legal authority to sign up rights agreement with the community.

(vi) The rights are legally justified.

(vii) There are accessible and fair avenues for seeking legal recourse in disputes.

In recent years however GoI has been sensitive to the issue of traditional and customary rights of forest dwellers and has launched the job of assigning land titles on an experimental basis to genuine persons having proof of long association with the forestland.

In order to smoothen the gradual transfer of rights and *de facto* conservation responsibility to concerned communities certain reforms need to be introduced in areas of forest management, regulations, monitoring and accountability system. Better forest mapping and identifying domiciled forest users, ensuring peoples' participation in planning approaches, matching local skills with use of local resources, safeguarding the interest of the forest dwellers from the clutches of middlemen, improving inventory system and channeling R and D to better adjust species diversity with livelihood needs are some of the important steps for a meaningful and efficient transfer of forest right to the really deserving community.

Current JFM endeavours in mobilising social capital and group formation has been inadequate and need to explore the potential of active participatory development. There exists a great deal of concern among tribal communities about the process of formation of JFM committees. Often they view the JFM as neglectful of existing and prudent use of natural resources, indigeneous knowledge and opinions and cultural issues. Villagers have often been found to ventilate their concern about the legal status of JFM committees which are subjected to dissolution by the Forest Department. There prevails a great deal of suspicion and uncertainty among villagers about the sharing of information, maintenance of minutes, records and accounts as well as timing and sharing of benefits from timber felling. Further in mixed villages, tribal people suffer from the threat of marginalised access due to restricting forest use among a defined group of people and their weak economic and social power. In order to establish a cohesive village society based on mutual trust and amity, it is necessary for developing a fully participatory micro-plan integrated with JFM efforts for each FPC area. This should focus on introducing various income and employment generating activities that might deliver the village people free from their dependence on forest resources. The FPC activities might be integrated with government-sponsored work programmes targeted to village masses through the efforts of the Forest Department. Together with grant of secured tenurial status, micro-planning of this type is very likely to lessen the extent of forest extraction and depletion.

References

Aihoon, J.K and J.K. Kirsten (1994). "A Review of the Environmental Impact of Agriculture", *Agrekon* 33:3, September.

Bedoya, Eduardo (1987). "Intensification and Degradation of the Agricultural Systems of the *Peruvian Upper Jungle*: The Upper Huallaga Case", in Little and Horowitz (ed.), *Lands and Risk in the Third World.* Boulder, Co: West View Press.

Burkard, G and S. Ebersberger (2002). "The Security of Land Tenure in Rain Forest Margin Areas: The Case of Central Sulawesi, Indonesia", Deutscher Tropentag, Oct. 9-11, Witzenhausen.

Hanna, S. and M. Munasinghe (1995). *Property Rights and the Environment.* Washington, The Beijer International Institute of Ecological Economics and the World Bank.

Molnar, A., S. Scherr and A. Khare (2004). "Who Conserves the World's Forests? Community—Driven Strategies to Protect Forests and Respect Rights", *Forest Trends Working Paper*, Washington D.C.

Roth, M. and D. Haase (1998). "Land Tenure Security and Agricultural Performance in Southern Africa", Southern Africa Program Publications and Outputs, *Broadening Access to Land Market*, (Online Web) *http://www.wisc.edu/ltc/baspubsafr.html#SA*